Border Lives

Women and Gender

THE MIDDLE EAST AND THE ISLAMIC WORLD

Editors

Susanne Dahlgren
Judith Tucker

Founding Editor

Margot Badran

VOLUME 16

The titles published in this series are listed at *brill.com/wg*

Border Lives

*An Ethnography of a Lebanese Town
in Changing Times*

By

Michelle Obeid

BRILL

LEIDEN | BOSTON

Cover illustration: The town of Arsal.

Library of Congress Cataloging-in-Publication Data

Names: Obeid, Michelle, author.
Title: Border lives : an ethnography of a Lebanese town in changing times / by Michelle Obeid.
Description: Leiden ; Boston : Brill, 2019. | Series: Women and gender, the Middle East and the Islamic world ; volume 16 | Includes bibliographical references and index.
Identifiers: LCCN 2019006550 (print) | LCCN 2019011372 (ebook) | ISBN 9789004394346 (ebook) | ISBN 9789004394339 (hardback :alk. paper)
Subjects: LCSH: Arsal (Lebanon)—Social conditions—20th century. | Arsal (Lebanon—Social conditions—21st century. | Women—Lebanon—Arsal—Social conditions—20th century. | Women—Lebanon—Arsal—Social conditions—21st century. | Ethnology—Lebanon—Arsal. | Lebanon—Boundaries—Syria. | Syria—Boundaries—Lebanon.
Classification: LCC HN659.A95 (ebook) | LCC HN659.A95 O24 2019 (print) | DDC 306.095692—dc23
LC record available at https://lccn.loc.gov/2019006550

Typeface for the Latin, Greek, and Cyrillic scripts: "Brill". See and download: brill.com/brill-typeface.

ISSN 1570-7628
ISBN 978-90-04-39433-9 (hardback)
ISBN 978-90-04-39434-6 (e-book)

Copyright 2019 by Koninklijke Brill NV, Leiden, The Netherlands.
Koninklijke Brill NV incorporates the imprints Brill, Brill Hes & De Graaf, Brill Nijhoff, Brill Rodopi, Brill Sense, Hotei Publishing, mentis Verlag, Verlag Ferdinand Schöningh and Wilhelm Fink Verlag.
All rights reserved. No part of this publication may be reproduced, translated, stored in a retrieval system, or transmitted in any form or by any means, electronic, mechanical, photocopying, recording or otherwise, without prior written permission from the publisher.
Authorization to photocopy items for internal or personal use is granted by Koninklijke Brill NV provided that the appropriate fees are paid directly to The Copyright Clearance Center, 222 Rosewood Drive, Suite 910, Danvers, MA 01923, USA. Fees are subject to change.

This book is printed on acid-free paper and produced in a sustainable manner.

To the 'Communist Sunni Muslim' who lived and died by his values

Contents

Acknowledgements IX
List of Figures XII
Notes on Names and Transliteration XIII

1 **Introduction: Border Lives in Changing Times** 1
  1 Figuring Out Border Lives 5
  2 Remoteness and Marginality at the Border 8
  3 The Ambivalence of Two States 12
  4 Rural Modernities 15

2 **Sociality between Movement and Space** 19
  1 New Capacities for Sociality 21
  2 The Workings of *'Ishra* 26
  3 Domestic Spaces, Gender and Consumption 30
     3.1 *House Interiors* 30
     3.2 *Gendered Sociality* 35

3 **Living Well: Experiments in Livelihoods** 40
  1 Livelihoods as an Ongoing Experiment 43
  2 Livelihoods in the Shadow of an 'Evil State' 47
  3 Contested Moral Economies 52
     3.1 *Living Well through Traditional Livelihoods* 53
     3.2 *Quarrying: Prosperity or Sentence to Hard Labour?* 56

4 **Pastoralists: Living the Past in the Present** 61
  1 Transhumance and Political Change 63
  2 Spatial and Human Organisation 66
  3 Herding Dilemmas 69
  4 Conflicts of Interest 71
  5 Envying 'the Comfortable Woman' 78

5 **Marriage between Love and Fate** 82
  1 The Befalling of Nasīb 84
  2 The Vocabulary of Modern Marriage 89
     2.1 *Love and Desire* 90
     2.2 *Partnership and Understanding* 92

    3 Intergenerational Negotiations   95
        3.1   *Zuhayr and Rima*   95
        3.2   *Zuhayr's Daughters*   98
    4 When Negotiation Fails   100

6 Suspicion and Scorpions: The Morality of Kinship   103
    1 Ensnaring Brothers and Suspicious Sisters   106
        1.1   *Female Altruism*   108
        1.2   *God Did Not Say 'Live Alone'*   110
        1.3   *Cinderella – But without the Prince*   112
    2 Of Failed Bargains   115
        2.1   *It is All the Mother's Fault*   117
    3 The Morality of Kinship   120

7 Local Elections: Politics at the Margin   123
    1 1963: Familism, a Divisive Force   126
    2 1998: *'Ā'ila* Redeemed   130
    3 Familism Strikes Back   132
    4 Corruption that Compromises National Pride   135
    5 The 2004 Lists: 'Old Wine, New Bottles?'   140
    6 New Council, New Directions   144

8 What the Future Hides   147
    1 A Visit in Post-Syrian Time   149
    2 Is it Possible to Move Backwards?   154

Bibliography   157
Index   169

Acknowledgements

This book is a culmination of a long journey, and I am indebted to each and every person who supported it on its way. I owe my deepest gratitude to the people of Arsal who always met those curious about their world with warmth, openness and a life-affirming sense of humour. My commitment to anonymity prevents me from naming the long list of dear friends who deserve to be thanked. I continue to cherish my lasting friendships with the individuals and families whose generosity and hospitality never cease to overwhelm me. Late night *mate* sessions, carpet workshop debates, big meals in various living rooms, long rides from Beirut to Arsal and exciting political debates on highland trips colour the writing of this book. I am honoured to have had the opportunity to share life with Arsalis and I hope that I have done them justice in my reconstruction of 'changing times.'

The work on this project started in the mid-1990s and would have been impossible without the support of several institutions. I am grateful to the Environment and Sustainable Development Unit (ESDU) at the American University of Beirut (AUB), particularly to Shadi Hamadeh whose commitment to a multidisciplinary approach made room for anthropological research and long-term collaborations; the International Development Research Centre (IDRC); the Population Council and Karim Rida Said Foundation. The generous grants of these institutions made field research and the initial writing up of this project possible. The Centre for the Advanced Studies of the Arab World's (CASAW) post-doctoral fellowship at the University of Manchester offered an invaluable intellectual space to write drafts of this book.

I am very grateful to the anonymous reviewers whose thorough reading and generous comments helped me improve my manuscript. At Brill, I thank Susanne Dahlgren and Judith Tucker, series editors for 'Women and Gender: The Middle East and the Islamic World,' for their patience and encouragement. Nienke Brienen-Moolenaar was extremely helpful in her guidance and I am grateful for her advice. Thanks to Peter Blore at the University of Manchester who helped me with the map and photos and I cannot thank Susan Kennedy enough for her careful edits of the manuscript and her infectious interest in the power of language. I owe special thanks to Portico Library. Its various cosy corners and warm and welcoming staff made the last and most challenging days of this project pleasant and hopeful.

Parts of chapter 2 of this book were published in 'Friendship, Kinship and Sociality in a Lebanese Town' in *The Ways of Friendship: An Anthropological Exploration*, edited by Amit Desai and Evan Killick, 93–113, by permission of Berghahn Books. A portion of chapter 4 was published in 'Uncertain Livelihoods: Challenges Facing Herding in a Lebanese Village' in *Nomadic Societies in the Middle East and North Africa – Entering the 21st Century*, edited by Dawn Chatty, 463–495, by permission of Brill. And parts of chapter 7 were published in 'The "Trials and Errors" of Politics: Municipal Elections at the Lebanese Border' in *Political and Legal Anthropology Review*, 34(2): 251–267.

Colleagues, friends, teachers and mentors shaped different versions of this work, though of course its shortcomings are solely my own. At the London School of Economics, I was patiently guided, encouraged, and challenged by Martha Mundy and Deborah James. The late Peter Loizos, Peter Gow, Laura Bear, Catherine Allerton, Michael Scott, Jonathan Parry and Chris Fuller were immensely helpful with ideas, suggestions and critique. I am grateful to Girish Daswani, Jason Sumich, Casey High, Evan Killick, Magnus Course, Maya Mayblin, Florent Giehmann, Eve Zucker, Giovanni Bochi, Carrie Heitmeyer, Bashir Bashir, James Brown and Yan Hinrichsen. Their feedback and stimulating conversations have influenced my project and its development more than they know. I owe deep gratitude to Amit Desai's friendship and intellectual companionship. He diligently read a full draft of the book and I am grateful for his invaluable insights that helped shape the final outcome.

My colleagues in the department of Social Anthropology at Manchester, who never tired of encouraging me to finish this book, have been an endless source of motivation. I feel extremely fortunate to be surrounded by a community of inspiring scholars who keep reminding me why anthropology is so exciting and why it matters. My thinking about kinship and gender has been thoroughly enriched by Jeanette Edwards' and Tony Simpson's work. Sarah Green, Stef Jansen and Madeleine Reeves broadened the way I approach questions of the state and borders. I especially thank Stef for his sharp comments on an earlier draft that he kindly offered to read and his constant encouragement. Angela Torresan, Soumhya Venkatesan and Atreyee Sen challenged me unreservedly, but always in friendly, positive and fun ways. For their heartening support, stimulating comments and intellectual generosity, I thank Pete Wade, Gillian Evans, Andrew Irving, Karen Sykes, John Gledhill, Lorenzo Ferrarini, Penny Harvey, Chika Watanabe, Petra Tjitske Kalshoven, Rupert Cox, Maia Green, Olga Ulturgasheva and Katie Smith. My gratitude also goes to Hoda El-Sadda and Dalia Mostafa in the department of Middle East Studies for their companionship and invigorating regional perspectives. In addition to intellectual stimulation, I thank Gillian, Angela and Rohan Jackson for sharing a much-needed wellbeing regime that kept me focused.

Several friends have left a mark on my intellectual journey in various ways. Sami Hermez, Mayssoun Sukarieh, Elizabeth Saleh and Youssef Chaitani have been a wonderful source of exciting conversation and intellectual guidance. I have learned so much from their work, knowledge and rich perspectives on Lebanon and have found comfort in their moral support. I owe profound gratitude to Zeina Misk and Sabine El Chamaa for their genuine interest and engagement with my work since the beginning of the project. I took great joy in their visits to Arsal and learned from their creative and rich insights. I thank them for their love and precious friendship. I also thank Abbas Boukhdoud and Amin Younes, for tirelessly listening to my outrageous stories over endless cups of '*makhbut*,' and for being unwavering friends over the years.

Finally, I would like to thank my family. No words can express my gratitude to my mother Hala, whose unconditional love, encouragement, and belief in me have been an essential anchor in every step of the way. Over the years, she has played an important role in my project as she forged friendships with my Arsali friends and followed the news of the town and its people with great care and affection. My sisters Rola and Rania are the rock of my life. Their resilience, enthusiasm and love are a constant source of strength. The culmination of this book is owed to their unreserved encouragement and their belief in the importance of pursuing one's passions. I am grateful to my brothers-in-law Faure Basulto and Ziwar Nakhesh, my uncle Jamal Khalaf, and my father-in-law Nicholas Kennedy, who have been incredibly supportive in different ways. Miguel, Kayan and Maxim, my gorgeous nephews, continue to bring joy to our lives. I thank them for the laughter, play and little words of wisdom that lighten up the greyest days.

More than anyone, my husband Ralph has endured my highs and lows as I struggled to complete this book. I thank him for standing by me without fail, for his intellectual honesty when I discussed my ideas with him, for his calming words when they mattered most and for his hilarious 'Rocky 1-inspired' motivational speeches that seemed to work more than I wish to admit.

Figures

1. Children having fun in the highlands   2
2. The expanding town of Arsal   4
3. Arsal at the northeastern Lebanese border with Syria   9
4. Women cooperating to grind dairy mixture (*kishk*)   26
5. An old mud-house: 'remnant of living heritage'   32
6. A woman preparing *mate*   38
7. Fruit orchards in the highlands   42
8. A family picking cherries   43
9. A herders' tent-house   68
10. Young shepherd bottle-feeding a lamb   72
11. Father of a bride dancing on the first night of the wedding   92
12. Bride and groom leaving the wedding ceremony to their new home   93
13. Protest against the American invasion of Iraq   139
14. Election leaflets for the Democratic Alliance   143

Notes on Names and Transliteration

Apart from names of political leaders and the town itself, I have changed the names of people that appear in this book as well as the names of their lineages. I have transcribed Arabic terms and expressions following standards established by the *International Journal of Middle Eastern Studies* (IJMES). When transcribing words from speech, I have maintained pronunciation and spelling of Modern Standard Arabic, except in instances when I use colloquial Arabic for certain terms, which I indicate in the text. I have not transliterated people's names and commonly known places (such as cities and regions). Throughout the book, translations from Arabic to English are my own.

CHAPTER 1

Introduction: Border Lives in Changing Times

It was a cold evening in February, and the roads were frozen and slippery. Um Hasan was excited to be paying a special visit to her long-time friend and relative, Um Abdu, to congratulate her husband Abu Abdu on completing his pilgrimage to Mekka, the *hajj*. So we huddled together, leaning on each other as we set out to the end of the road, taking tiny steps in the dark, narrowing alley, careful not to slip. Um Abdu greeted us and led us to her living room, where a tray of perfumes, prayer beads and little bottles of Zamzam holy water adorned the entrance. The familiar *hajj* souvenirs would be distributed to us and other well-wishers on our way out. Abu Abdu was not yet home and his family was entertaining visitors. We sat with Um Abdu, her sister, three daughters, and two neighbours, chatting and sipping freshly brewed tea as we sampled the special Saudi dates. One of Um Abdu's daughters was lying down on a corner mattress, wrapped in winter scarves and pale with a cold that seemed all too common in the neighbourhood that freezing month. 'Look at her, feigning death like her mother!' teased Um Hasan. Her comment sparked chuckles as it evoked the memory of Um Abdu's mischievous youth and offered an invitation to tell a story. Apparently, Um Hasan and her friend were on an errand to fetch water in the *jurd* (highlands) when they stumbled on Um Abdu, lying down alone, feigning death. 'She didn't move or even breathe!' commented the neighbour who had heard the story before and seemed ready to own the details as if she was there herself. 'Um Hasan was so affected by this catastrophe that she dropped to her knees and began to chant a *'atāba* (a song of loss and lament).' Her passionate mourning, however, came to an abrupt end when Um Abdu surprised her by rising from the dead and joining in the singing! All three women were now interrupting each other with details as they uproariously recounted Um Hasan's reaction; 'The look on her face!' The daughters listened attentively, amused to hear funny stories about their mother, now in her late 50s. Um Abdu laughed tearfully at her own prank as she revealed how she had carried out her plan. When the laughter settled, Um Hasan sighed. 'Back then, we used to laugh from the bottom of our hearts.' 'Oh yes!' Um Abdu agreed. 'Life was more beautiful in the old days. These days, there is nothing to laugh about. And if we laugh, laughter seems to have no savour.' A mood of reflection descended on the room. 'Everything changes,' commented Um Hasan, 'very much so. But in this life, one must *yu-maddī*' – an expression that literally means 'make the present past,' or simply 'live on' or 'pass the time.'

FIGURE 1   Children having fun in the highlands

There is ambiguity in the way transformation in social life is expressed in the northeastern Lebanese town of Arsal. Um Hasan and her friends' romantic depiction of a life gone by was a common way of recalling the not so distant past among the families that I worked with. Such takes on the past occupied many a storytelling evening during *ziyārāt* (social visits), a common pastime in rural Lebanon, especially in places where commercial recreational and leisure outlets are scarce. These stories tended to conjure up, perhaps selectively, images of innocence and lightheartedness, laughter and naivety, of childhoods enabled by living with nature as people dwelled in the highlands for long periods of the year. The richness of social life, people claimed, softened the harsh conditions of everyday pastoralist life in the highlands. My interlocutors reconstructed a time when they had no means, when there was poverty, but when people's needs appeared humble.

> All we needed was a bite to eat. You should have seen how they dressed us when we were young. Boys or girls ... barefooted, we wore torn *abāyas* with nothing underneath ... you could feel the wind whistle under your dress. They were different times. People back then just lived, but we were happy.

The way we choose to talk about the past often relates to how we perceive the present. The exaggerated simplicity depicted in a common version of a past in which people 'just lived' without the demands of modern day life is contrasted with people's perceptions of life in the present as being too complex (*muʿaqqadda*). Um Hasan and her generational cohort often commented on how quickly the village (*dayʿa*) they grew up in had changed. 'Arsal is a city now,' people would remark in reference to the physical expansion of concrete into the hills surrounding the valley and the dense population that inhabited it. They would ponder over how this expansion had affected social relationships in the village that had become transformed into one of the largest towns of Lebanon's Northern Biqaʿ: 'Today, no one knows anyone anymore,' people would say to convey the loss of a sense of familiarity. Yet, it was not uncommon to hear an altogether contradictory appraisal of how much life had improved for people. Um Hasan herself would often say that the younger generation was more comfortable (*murtahīn*) than hers and previous generations ever were, that people today were generally more empathetic and forgiving than they were in the past and that her own daughters had opportunities – to learn, to work, to choose their husbands, to travel – that would have been unprecedented and unthought-of in the family she grew up in. These contradictions were at the heart of figuring out how times were changing.

The metamorphosis of Arsali sociality was omnipresent in the everyday narratives and (inter)actions I witnessed over the years. Mundane conversations, such as children bargaining for a new set of clothes for *Īd* (the Muslim holidays), would prompt a tirade that began with 'In our days …' from parents and grandparents about the differences they perceived in time and place. *Al-qadīm* (the old days) became a regular turn of phrase that contrasted with *al-yawm* (today). Whether to critique the present (or past) or to celebrate it, these expressions were used, sometimes even appropriated by young people, to comment on the drastic changes that had taken place in dress, gender relations, education, kinship, literacy, consumption, political transformations in the town and the country, and the rise of various technologies that increasingly connected Arsal to the outside world and its influences. To be sure, this 'temporal consciousness' (Limbert 2010) permeated everyday life, in which the past resurfaced to shape, shift, endorse, and contest the present.

Notwithstanding the Lebanese war of 1975–1990 (known as the Civil War)[1] and the havoc its long-lasting repercussions wreaked on the entire country, this

---

1 Lara Deeb choses the term 'uncivil war(s)' to refer to the internationalization of the wars in Lebanon and to highlight that 'no war is civil' (2006: 66; footnote 1). Taking both these points on board, and particularly the contentious internationalization of these wars, in this book I

FIGURE 2   The expanding town of Arsal

period brought great change to rural economies on the northeastern border. The introduction of fruit farming to a semi-nomadic agropastoral way of life, together with the creation of an extractive industry in newly exploited quarries in the highlands, altered modes of living. This was visible through changes in land-use, in mobility patterns (in Lebanon and across the Lebanese/Syrian border), in living arrangements between the highlands and the town, and in the spatial reordering of the neighbourhood. Together, these created vicissitudes of sociality in the town as Arsalis experienced significant changes in their rural space – the 'layers of social, political and economic relations that traverse different physical spaces at any one time' (Marsden 1998: 16).

This book is concerned with how people in post-war Arsal lived, in Um Hasan's words, '*yu-maddū*' (sgl. *yu-maddī*), in 'changing times.' It asks what the 'passing of time' entails in rural lives perceived as being fraught with obstacles and hardships. Over the years I spent with the generous Arsali families that befriended and hosted me, I frequently heard a well-used mantra being invoked to remind listeners of the local philosophical basics of living: that 'life was but

choose to specify the wars by referring to their years rather than accepting the problematic category 'civil war.'

a couple of days [long] that needed to pass, whether in goodness or in ugliness' (*hal yawmayn bidhum yamdū, bil mnīh aw bil dʿīr*). The underlying message of this saying is that in between goodness and ugliness exists a space for improvisation. 'Changing times' unsettled social life and demanded new ways of thinking, acting, and being; they required navigating, if not embracing, ambiguity. One way or the other, it was suggested, humans needed to work things through and work them out. In seeking to document post-war Arsali lives, I therefore deploy the local expression of *yu-maddī / yu-maddū* to explore the everyday labours that people exerted in and over time, within the limits and conditions of a hard life, so that they could 'get on with things.' The chapters of this book explore the social processes, institutions, and practices that my interlocutors felt have changed most since the outbreak of the 1975–1990 war: their physical environment and social space (Chapter Two), their traditional livelihoods and household regimes (Chapters Three and Four), their spheres of intimacy (Chapters Five and Six) and their political organisation (Chapter Seven). In my discussions, I pay attention to the way people from different generations 'figured out' how to live their lives in time of transformation. Getting on, as the chapters that follow will show, takes different shapes – of people playing along, experimenting, enduring, laughing at life, rebelling and resigning themselves to fate – as they figure things out. Given that both past *and* present are inscribed within the expression *yu-maddī* (hence 'to pass the time in the present'), I observe the ethnographically emergent contrasts and continuities between 'the old days' and 'today' in their constant conversation.

1 Figuring Out Border Lives

Consideration of my interlocutors' everyday trials in 'figuring out' has tended to raise problems of my own in 'figuring out' what have proved to be some key methodological challenges looming over this book. How does one deal with problems of scale when writing about a marginal place like Arsal? While the experiences of transformation may seem much more profound in Arsali social life than in its urban counterparts – though of course people *everywhere* talk of social change across generations – these changes cannot be conceived in isolation to the national and regional contexts that create the conditions in which rural change occurs in Lebanon and elsewhere. As early as the build-up to the war, Lebanon suffered 'rural disintegration' (Nasr 1978) under the hegemony of financial and commercial sectors that privileged the capital and fostered regional inequalities. During the war, protracted violence reverberated in diverse ways across all of Lebanon's landscapes. Although Arsal was not a frontline town per se, the militarised use of its vast highlands and the roles its residents played in the war point to involvement in events that go beyond the marginal. As Lebanon entered its post-war period, neoliberal approaches to reconstruction and 'peace' favoured, once again, the centre at the expense of Lebanon's peripheries. These, along with the changing relationship between the Syrian and Lebanese states (and the successive governments in Lebanon that reflected the schisms that ensued), turned the border into a geopolitical conjuncture that is continually shifting. The story of change in Arsal is therefore historically contingent and deeply entangled with the economic and political changes of the country and the region. These scales are not mere context or background; they materialise in everyday life. How, then, can we capture all these elements of scale without losing sight of either everyday experiences or the larger picture? My ethnography traverses different spaces of Arsali life as I interlace large-scale socio-economic and political changes at the border with people's intimate lives. Shifts in livelihoods are inseparable from state-society relations and the specific regional history of the northeastern border of Lebanon with Syria. Gendered personhoods, new ideologies about domesticity, piety, love and marriage are co-constituted with these same shifts, as are reconfigurations of family, household, and the moralities that bind people together. These entanglements, and the manner in which they played out in the everyday, became the interest of my research.

The other question that preoccupied this project was how to write about a long-term perspective such as the one the book tries to cover. The question of time is as central to my interlocutors as it is to me, the anthropologist, particularly since I have known the town for over 20 years. While incredibly enriching, this long-term knowledge can be elusive and a challenge to the task of creating a coherent narrative. One difficulty confronting me was from which point in time would I best be able to capture the tempo and scale of 'changing times' that seemed to occupy my interlocutors? For example, what kind of story might I have written eight or fifteen years ago? Ironically, during the years I was based in Arsal (2002–2004), my interlocutors commonly depicted their town as a slow-paced place that never changed. When friends from the city returned for visits and asked them the usual 'What's new in the town?' they sheepishly responded *'ma fī jadīd fī Arsal!'* There was no such thing in Arsal; it was not the kind of place where new things happened. Yet, mundane conversations and interactions in these times revealed the textures of change and the ways it steered people's lives. Conversely, during elections, or national events that shook the country politically, these same interlocutors would claim instability and flux as fundamental attributes of border living. While the book incorporates stories that date back to the 1950s, I focus on a period of 15 years

(1995–2010) that starts shortly after the end of the war and ends roughly before the outbreak of the Syrian protests in 2011, events that continue to change the border (and the two countries) in profound ways. The knowledge and material I present here is constructed in retrospect, with the advantage of time that has allowed me to situate change in the mundane with a wider temporal purview, even when I limit the narrative to particular instances and periods.

Ethnographic fieldwork, particularly during 13 months between 2002 and 2004, and numerous regular visits over the years have allowed me to follow the trajectories of the lives that I recount in this book. The families that I write about belong to different lineages. They had various income and educational levels and lived in different parts of the town. In spite of their differences, they shared a common pastoralist background. In all of them, the eldest generation grew up in herding families and experienced seasonal movement between Lebanon and Syria. Their children and grandchildren experienced very different trajectories, with the growing importance of education, the increase in job prospects, and new post-war opportunities. Only one of these extended families maintained this regime during my fieldwork years. In the chapters that follow, I discuss some of the intergenerational tensions experienced in its households as the younger generation resisted herding and their forefathers' lifestyle in the context of alternative livelihoods and new consumption regimes. In Arsal, families tended to live in multi-generational households, in the same building structure, usually two-story houses, or in compounds of houses. Through my immersion in their social worlds, I experienced the rhythms and socialities of the different neighbourhoods (*harāt*) in which these families dwelled.

The families and individuals that appear in this book are in one way or another linked to a rural development NGO to which I was affiliated through a multidisciplinary action research project (Obeid 2012) led by the American University of Beirut (AUB) and in which I had rented a room. Given its successful profile in the town, with a multiple-lineage membership that, at its height, reached 200 members, the NGO acted as the project's 'local facilitator.' It hosted a permanent carpet-weaving workshop and supported a Women's Food Cooperative and a Herders' Cooperative, both founded in the early 2000s, when the NGO had become a social hub. Thanks to its large hall, busy kitchen, and hospitable members, the NGO became an inclusive, open meeting space for members and non-members long after working hours. Punters would drop by to sip tea with the volunteers. The women at the workshop lunched together, often cooking meals on the day and inviting whoever was on the premises to join them. I spent long hours with the 'carpet girls' (*banāt al-sijjād*), as they were known. I joined them for tea breaks and watched them make beautiful prints on their large looms as they talked, joked, bickered, and debated social issues. It was through my daily after-work visits to their families that I was able to observe and participate in their everyday life.

To reconcile questions of scale and time, I chose to explore *border lives* as an account of the place and people of Arsal. When I began my fieldwork, I had not envisioned studying the border in any direct manner. However, attention to people's lives prompted me to appreciate the central role the border played (and still plays) in Arsal and the extent to which the social, economic and political lives of Arsalis depend on its porousness. The book therefore utilises 'the border' as a spatio-temporal lens that reveals to us how lives lived around and across it are transformed in short periods of time (Green 2010; Radu 2010; Reeves 2014). We can thus appreciate how an unexpected arrest at a shifting border might profoundly change the life course of a family. Or how the loosening and tightening of security and bureaucratic measures enhance and restrict livelihoods that depend on border crossing, thus impinging on labour arrangements, gender roles, and relations as well as kinship. We can also trace how political ideologies on both sides of the border were contested and appropriated at very local levels. In this sense, I see the border as a living and transformative space that shifts and leaks; it 'hold[s] varying sorts of meaning for different people' (Migdal 2004: 5). This is particularly the case at the Lebanese/Syrian border – an 'invention' according to Elizabeth Picard (2006: 76) – which has always been characterised by ambivalence.[2]

2   Remoteness and Marginality at the Border

Tucked on the slopes of the Anti-Lebanon Mountains, 80 kilometres north of the capital Beirut, the town of Arsal stretches over 475 square kilometres of highlands (Hamadeh et al. 2006: 1), which it shares with other villages that border Syria from the southwest. Narrow unpaved roads, orchards dotting the hills, and chickens strolling around front yards gave the town I first encountered its stereotypical air of rurality. Arsal had one of the largest livestock counts in the country in the 1990s.[3] Most of the older generation in the town had been raised in families that followed a traditional form of semi-nomadic agropastoralism, the main form of subsistence up until the second half of last century. By the end of the century, residents had moved to a mixed economy that incorporated crop production. Yet the scene that welcomes the visitor today, soon after driving through the village of Labweh in the valley, presents an image very

---

2 This is the case even with a securitized border such as the 'Blue Line' at the Lebanese-Israeli border. See for example Beydoun (1992), Meier (2013) and Schneidleder (2017).
3 The estimate is at 200,000 goats and 500,000 sheep (Hamadeh 2002).

INTRODUCTION: BORDER LIVES IN CHANGING TIMES 9

FIGURE 3   Arsal at the northeastern Lebanese border with Syria

unlike that of an agricultural haven. The main road runs between a gaping hole in the mountain on one side and a locally owned quarry, with its equipment, machinery and stacks of rocks and tiles, on the other. We are in the presence of an extractive economy, the scale of which becomes clear as one goes deeper into the highlands. A mass of concrete emerges from the valley and stretches out to the horizon with half-built houses expanding into the hills. These material contrasts seem to evoke the quality and pace of change in the town.

On my very first visit, the bus driver explained to me that we were heading towards a 'remote' area (*nā'iya*). The town lies in the Northern Biqaʻ region, one of Lebanon's marginalised areas (*muhammasha*). Along with the 'South' and the 'North' (Volk 2009), the Biqaʿ constitutes a periphery characterised by economic disenfranchisement, politics of neglect, and state abandonment. I later learned that this description was used both by people living in the capital, who tend to think of the Northern Biqaʿ as an economic and cultural periphery, and by the residents of the town who experience remoteness from their vantage point. Remoteness is not just about perceived geographical distance. The term is charged with a sense of economic and cultural delay. On my earlier visits, I was told in jest that the original ancestor of the Arsalis must have been a run-away thief who had no choice but to hide among the hills, which are a 30-minute drive from the main road. 'Why else would one want to settle so far away from society?' In self-deprecating tones, the men and women that I spoke with described their town as being 'outside history and geography.' This spatio-temporal expression spoke volumes of what it means to live on the margin of state and society. It was not ruin or carnage that signalled a civil war had come to pass in the town; rather, the war was present in the continued neglect and abandonment by a state that was too slow and selective in its recovery and had left the town lacking in basic services. The only sign of state presence in Arsal in the early post-war years was a local police station run by two officers (*gendarmes*) who were tasked with the security of a population estimated at 32,000.[4] In that period, the local municipality was immobilized and, like many municipalities across Lebanon, was only reinstated in the first post-war local elections in 1998. This meant that whole infrastructures were left unattended: roads (that connected the town to the highlands and to the valley), water, sewage, electricity, and waste management. The town had two state-run schools, primary and secondary, and ten privately owned ones. With limited places available, many families had to enrol their children in mainly Shiʿi schools in the valley, a controversial option in the context of a rising post-war Sunni identity among residents. Medical services were noticeably scarce, with just one clinic that housed two nurses. I knew of less than a handful of local nurses who worked outside the town and conducted house visits in their spare time. But to get to a hospital in an emergency, residents had to reach the nearest city of Baalbak, about 45 kilometres away, across unlit, bumpy, and eroded roads. With this inaccessibility of medical care, and the exorbitant prices charged by private doctors in Lebanon, the town's residents relied on the open border in order to seek medical advice, among other services, in the Syrian city of Homs. Through the official border in the far northern town of Qāʿ, residents made

---

4 This is the Municipality estimate, as I learnt from an interview with its main clerk.

INTRODUCTION: BORDER LIVES IN CHANGING TIMES

regular day excursions to Syria. The frequency of these trips was encouraged by a blasé approach to security checks at that border checkpoint.[5]

One particular service brought home the predicament of being even further behind than their neighbours in the Northern Biqaʿ. In the mid to late 1990s, the national daily newspapers reached Arsal a day late. 'We are reading stale news (*akhbār bā'ita*)!' my interlocutors protested. I became accustomed to the deployment of humour and hyperbole to comment on social issues in the late evening *ziyārāt*.

> In the 1970s, when the first *makana* (car) drove down the hill, it was a big thing here in Arsal! People had not seen anything like it! Poor Um Ali! She yelled in panic that a giant bug was attacking the village!

Such an anecdote of self-ridicule would be readily met with another:

> When radios were introduced to Arsal, one man switched it on to the BBC (Arabic service). The broadcaster announced in Arabic, '*huna London*' (this is London). The Arsali threw the radio away angrily and corrected, 'no! This is Arsal!'

And another:

> My neighbour, the *khityāra* (elderly woman), went to Beirut for the first time in her life. When she returned from her visit, she went around telling her neighbours that dogs in Beirut understood French! She swore that the dog responded to its owner when it heard her say '*Vien!*' (Come!).

These caricatured sketches merited their humour only when framed *in relation* to other parts of the country. The contrast presented was stark and intended to make an impression on the listener. The unequal relation they portrayed was between a Lebanon that boasted its 'golden age' before the 1975–1990 war and the Arsalis who were only just being introduced to cars and radios at that time. As the capital underwent large-scale, state-sponsored development and reconstruction unmatched in any other region of Lebanon in the aftermath of the war, the Arsalis were still behind the times, reading old news. These other places were imagined to be more glamorous and advanced, hosting cosmopolitan

---

5 There are two other international border-crossing points to Syria. The main and busiest one is in Masnaʿ in central Biqaʿ. The other, ʿArida, is a smaller crossing point near the northern city of Tripoli.

wealthy elites who led bourgeois lives and spoke foreign languages to their pets. Marginality, as Sarah Green argues, 'explicitly evokes a sense of unequal location as well as unequal relations ... [and] a difficult and ambivalent relevance to the heart of things' (2005: 1).

Although marginality is a predicament of the Northern Biqaʿ as a region, one might borrow from Michael Gilsenan to describe Arsal as a 'periphery of a Lebanese periphery' (1996: xi), when we consider the town's political positioning vis-à-vis regional politics. Arsal is the largest Muslim Sunni town in an area that has a majority Shiʿi population (there are some Christian and mixed villages in the area, but they are a minority). This alone, of course, does not explain inter-regional tensions, but it does shed light on the manner in which local inter-village differences were woven into macro-political processes over the years. At the heart of the unequal relations that developed on the border after Lebanon's Independence in 1943, lies the fluctuating relationship between the Syrian and Lebanese states and the loyalties they invited (and inhibited) on the ground.

3   The Ambivalence of Two States

As early as the first Lebanese civil war (1958), the border played a central part in the political positioning of Arsal nationally as it shaped the residents' relationship with both their Biqaʿi and their cross-border Syrian neighbours. Arsalis in general supported the United Arab Republic project, leading them to take part in the rebellion against President Kamil Chamun's rule (1952–1958) in what turned into a short-lived civil war.[6] They received armed assistance from the Syrians across the unofficial border, which was 'indescribable' (Reeves 2014: 3) in the sense that there was no clear territorial division between the Lebanese and Syrian states, eventually giving them victory over their neighbouring Lebanese opponents in the valley who were state supporters. This small success, although it won them a reputation for fierceness in the Biqaʿ, came at the cost of a brutal reaction from the Lebanese state air force, which bombed the town for hours on end as residents fled from their fields to hide. These events were crucial in defining the peripheral status of Arsal in the northern Biqaʿ. The town experienced a prolonged period of state neglect after these events (Obeid 2010b). State antagonism, however, was de-politicised by emphasising stereotypes associated with Lebanon's 'frontiers' (Watts 1992): 'particular

---

6 See Salibi (1988) for a historical analysis of identity formation in Lebanon. For details on the first civil war, see Hanf (1993), Picard (2002), El-Solh (2004) and Traboulsi (2007).

kinds of places' that are considered 'savage, primitive and unregulated' (Volk 2009: 264; Das and Pool 2004). In the meantime, Arsalis enjoyed open access to the Syrian border, its residents relying on Syrian villages and towns as well as Syria's pastures in the lowlands for their everyday dealings and livelihoods.

This cross-border affinity began to change in the course of the second Lebanese civil war (1975–1990). Unlike other parts of the country, such as Mount Lebanon or the capital, the town itself was not fought over, but its vast highlands became a training ground for the militarised political parties that proliferated in the town and recruited locals: Communists, Arab Nationalists, Syrian and Iraqi Ba'thists, and several Palestinian factions became active in Arsal.[7] By the early 1980s, many Arsali men (and a few women) were moving around the country with their political parties. Towards the end of the war, Arsal had made its mark nationally due to its active participation in fighting Israeli forces during the 1982 invasion. Some 250 martyrs are said to have died, a noteworthy number that awarded the town the epitaph *Umm al-Shuhadā'* (Mother of Martyrs), at least among left-leaning circles. As some of the following chapters will show, this mobility through political party activism led to the circulation of new ideas, things, relationships and outlooks that reverberated in the social life of the town. Amidst these transformations, the relationship with the Syrians, who in the 1950s were Arsal's allies in their quest for Arabism, started taking a different shape, *politically*. After the war broke out, Syrian troops entered Lebanon in 1976. Although this move was to 'save the integrity of Lebanon,' as the late President Hafiz Al-Asad promised (Picard 2002: 190),[8] it marked the moment when Lebanon – and especially the Biqaʿ – came under the control of Syrian state apparatuses, a control that lasted right through till 2005. This period saw the birth of a new kind of antagonism towards Syria in Arsal, and generally in other areas in Lebanon. Economically and socially, by contrast, the border remained open as people still crossed to visit Syrian markets, to call on relatives, and to camp in the lowlands.

The architecture of the agreements that led to the end of the 1975–1990 war placed the northeastern border of Lebanon, and Arsal in particular, in a peculiar predicament. The 'Document of National Understanding,' also known

---

7 See El-Khazen (2003) for a discussion of the growth of the political left in Lebanon.
8 Syria's presence in Lebanon had strategic interests: After Israel annexed the Golan Heights in 1967, Syria's President Asad believed that 'Lebanese and Syrian security were interdependent' (Picard 2002: 114). The protection and defence of Syrian territory presupposed military control of the Biqaʿ valley which 'forms a natural corridor of access to the central Syrian cities of Damascus, Homs and Hama for an Israeli army that had gained foothold in a southern Lebanon' (ibid.).

as The Ṭāʾif Accord,[9] is considered to have ended the conflict in Lebanon. However, it did *not* rule out Syria's involvement in post-war Lebanon. Instead, the Presidents of the two countries signed a 'Brotherhood, Cooperation and Coordination Agreement' in 1991 that, if anything, legitimized Syrian control over Lebanon, despite the stated intention and plan for Syrian troops to withdraw from the Biqaʿ as early as 1992 (Traboulsi 2007). In most Lebanese areas, especially the capital, the Syrian military presence gradually declined over the years as troops withdrew to a number of strategic bases and handed over security to Lebanese apparatuses, which still collaborated with the Syrians (Picard 2002), at least in the first decade after the war. In the Northern Biqaʿ, on the other hand, the Syrian presence was less discreet, with several permanent checkpoints planted on the main road and a high degree of control and intervention in local affairs. Residents of the northern Biqaʿ had no illusions about who controlled them. Their lives were monitored by the figure of a Syrian *muqaddam* (army officer) and the Syrian *mukhābarāt* (intelligence services). Where the state still felt absent after the war, Hizbullah was the actual power in the Northern Biqaʿ and enjoyed a relatively harmonious relationship with the Syrian state.[10]

Arsali residents' reliance on Syria in the post-war period was ambivalent as they felt caught between their historical social and economic ties with their cross-border neighbours and the political apprehension they felt towards Syria, a state that threatened their nationalism and sovereignty (Obeid 2010b). As Lebanon fell under Syrian control, it was the border where lack of sovereignty was most intensely felt.[11] For, in spite of the rhetoric about the weak, fractured, absent or failed Lebanese state, the state *did* function, though in some places more than others (Volk 2009; Hermez 2015, Kosmatopoulos 2011; Mouawad & Bauman 2017; Obeid 2015). To the people I worked with, the fact that Syrian cities were more accessible to the Biqaʿ's inhabitants than nearby Lebanese ones consolidated the status of the region as a margin.

Arsal had twice won a seat in Parliament in the post-war years, but both MPs, who resided in the capital, were believed to have been unsuccessful in bringing any improvement to the town. As scholarly studies of Lebanon repeatedly

---

9   The Ṭāʾif Accord was signed in the city of Ṭāʾif in Saudi Arabia under Saudi and Syrian auspices. See Krayem (2003) and Norton (1991).
10 See Norton (2007) for more details on the relationship and tensions within Hizbullah leaders (especially Sobhi Tufaily in the Biqaʿ) and their conflicting views on 'entering politics' in the early 1990s.
11 This is characteristic of other borders in the region in which 'respect for international boundaries and territorial sovereignty' is not 'systematically an operational principle in relations between states' (Brandell 2006: 16).

inform us, citizenship codes and practices are embedded in patrilineal and patriarchal rules as well as sectarianism (Joseph 1999; Johnson 2001). 'Lebanese citizens need brokerage, *wasta* (personal connections), to gain access to services and resources' (Joseph 1999: 311). This was remarkably lacking in postwar Arsal. Despite Arsal's large Sunni constituency, the residents did not have a political leader who would act as a patron, a requisite in the Lebanese clientelist political system. In the immediate years after the war, national Sunni leaders like Prime Minister Rafiq Hariri, who gained popularity in the town only after his assassination in 2005, were criticised within Arsali circles. Hariri's development plans were infamous for their 'social coldness' (Perthes 1997: 17; Baumann 2017) and regional inequality. They privileged reconstruction of the capital's centre over other areas and invested in economic infrastructures while increasing domestic and international debt. In the peripheral regions of Lebanon, state neglect was tangible in the lack of such investment and in poor basic services. Yet, in the Northern Biqaʿ (and the South), Hizbullah filled gaps left by the state and invested in social services through building hospitals, schools, medical centres and a variety of operative NGOs. Hizbullah had no presence in Arsal, however, and offered no such services. This unequal relation heightened the sense of marginalisation among its residents.

A sanguine outlook seemed to mark the first period of the post-war years when people 'yearned for normal lives' (Jansen 2014: 241) and tolerated the frustrations generated by the slow pace of institutional recovery. But that period was also fraught with the tensions brewing in the country as a result of unequal development and reconstruction, internal divisions in government, and the looming question of the Syrian presence. These dynamics were unfolding in distinct ways at the border and had a bearing on the socio-spatial lives of its residents. The following chapters make it clear that decisions that seemed very 'domestic,' or even intimate, were inextricably linked to certain economic regimes in the town, to the politics of the border, and to the presence or absence of the Lebanese and Syrian states. These exacerbated the experience of remoteness from the centre and reinforced marginality as a condition of the border, thus leading to novel ways of 'working out' how to navigate the ambivalence created by this post-war predicament.

4 Rural Modernities

Arsalis referred to themselves and to the surrounding Northern Biqaʿ as a rural society (*mujtamaʿ rīfī*). What 'the rural' constitutes is by no means static; it shifts in meaning and value. Some practices, customs, and traditions were considered more 'rural' than others. Being and belonging to the rural was something to be proud of, sometimes. At other times, it was used to explain frustrations with the town and its residents and to express the slow pace of change. There exists a persistent view that the rural is distinctive from the urban. Rural sociologists and geographers have grappled with what exactly defines these distinctions, particularly in parts of the world where agriculture is not the only marker of the rural and where we can no longer gauge the urban only through industrial development (Mormont 1990). In Lebanon, the rural and the urban are interconnected in many ways, not least through the scale of migration, displacement, and return migration that has occurred between villages and cities during several periods of violence (El Nour et al. 2015). The rural tends to evoke nostalgia for 'nature,' beautiful landscapes, and a slower and better quality of life characterised by generosity, hospitality, and rich social relationships. It invokes 'roots,' given that many urban dwellers originate from villages and maintain ties by visiting their villages on weekends and by making plans to retire there even when they live in cities. At the same time, the rural, especially in Lebanon's peripheries, is imagined to exist in 'a time warp' (Volk 2009: 265). Urban elites and liberal modernists, including development agents, have caricatured rural inhabitants as passive agents and generally as 'less enlightened subjects' who need to be taught 'empowerment' (Abu Lughod 2009: 91). In her work on modernity, Lara Deeb notes that 'as the modern unravels, it becomes about comparisons, boundaries between groups, relations of power, identity, similitude and difference' (Deeb 2006: 16).

Most people I talked with agreed that the *rīf* (countryside) and Arsali society needed to develop. For many, progress was about access to services, infrastructures, and technologies. But it was also about 'human development' sought through increased awareness (*wa'y*) and knowledge. Indeed, the value of human development was frequently expressed as a criticism of Prime Minister Hariri's approach to post-war development. He was seen to 'privilege *hajar* (concrete) over *bashar* (people)'. Within the town, societal progress was articulated in the language of development (*tanmiya*) that became the currency of new NGOs emerging in the early post-war years (Obeid 2012). Whether it was *tatawwur* (progress) or *tanmiya*, these idioms of modernization where used in a positive light. Modern-ness, however, was a little less straightforward.[12]

In Arsal the identification of 'rural' captured a shared way of life in the region, one that somewhat differed from urban life. This reference and relation to the urban was ambiguous. I often heard people tease or criticise someone

---

12 See Lara Deeb's (2006) Introduction for an excellent discussion of the distinction between these terms.

for being *mutamaddin/a* (urban[ised]) when they failed to do things in a 'rural' way. For example, someone who falls short of living up to expectations of hospitality by not offering a meal to guests coming from far away might be accused of being *mutamaddin/a* (sometimes *moderne*, in French, was also used). I heard my friend chastise her married sister for not wanting to help her and her mother to prepare their year's supply of pickled aubergines, but still expecting a share. 'You have become *mutamaddina*!' she reprimanded her sister, 'You don't get your hands dirty anymore.' Refusing to take part in agricultural activities, stretching the boundaries of prescribed gender roles, favouring the self over the family, or flaunting consumption styles that were not common in the town were all taken as indicators of *tamaddun* (urbanisation). To an extent, *tamaddun* seemed to threaten 'the rural' and to erode certain values, practices, and moralities. Yet, there was a certain desire and attraction for the urban. I am of course by no means advocating that we should accept that the rural and the urban (and by the same token tradition and modernity) are dichotomies. Rather, I am suggesting that ethnographic attention to these tensions and contradictions in which people grapple with boundaries and thresholds of 'the old' and 'today,' 'the rural' and 'the urban,' and 'tradition' and 'modernity' are at the heart of conceptions and embodiments of what it means, and what it takes, to be modern. They also shed light on the uncertainties surrounding the manner and pace in which the rural is changing. Rural modernities are experienced in multiple ways. There is plenty of room to discuss what constitutes the rural in Lebanon and the Arab world today and to animate the vagaries of rural social life and its transformations.

In spite of the critique of these dichotomies and the idea that the urban and the rural constitute 'two worlds' (Rigg 2014), people inhabiting rural areas often maintain that their lives *are* different. The rural is often construed through a 'particular cultural and moral milieu that ... makes rural people distinctive – their concern for family and community, their moral economy of sharing and communal support, their conservativeness, and their self-reliance and dislocation from the mainstream' (Rigg 2014: 3). The people I worked with may well agree with this description. But the universalism of this depiction undermines how the countryside transforms, how these very social formations (family, community, moral economy etc.) are (re)fashioned whether through education, mobility, changing landscapes, national politics, transforming economies or the omnipresent *desire* for modernity. It also undermines how different groups and individuals within rural contexts distinguish themselves from co-residents they see as more 'backward,' 'conservative,' or 'modern,' precisely as a result of how they engage with family, gender, community, and economy. Furthermore, such blanket descriptions of the rural underestimate the extent to which these social and moral formations are valued and championed by urban dwellers. For example, the modernist idea that the urban family nests or (ought to) in the private/domestic domain as it is stripped of its political and economic functions does not ring true in the Lebanese context. Suad Joseph's work, in particular, has shown that the family 'lies at the core of society – in political, economic, social and religious terms' (1996: 194), regardless of region or location. As Susan McKinnon and Fenella Cannell have argued, 'the narratives we tell ourselves about how modern social life is different from, and differently structured than the past ... [is but a] *myth* of modernity' (2013: 8; original emphasis).

As I explore spaces of social change in the book, I pay attention to new and changing metaphors of modernity. My aim is to recast the rural as a site in which we can investigate the modern. By investigating lives at Lebanon's northeastern border, *Border Lives* hopes to fill a gap in the emerging rich ethnography of Lebanon (and the Arab region) that has tended to be predominantly urban-centric.[13] In shifting our attention to areas considered to be 'marginal,' the book aims to re-centre both the border and the rural into conversations about modernity and social change.

---

13 A wealth of studies on urban settings such as Cairo, Beirut and Dubai, among others, has emerged in the last few years. Deeb and Winegar (2012) argue that this shift to urban studies has helped replace images of the region as being tribal or exotic. But it has also shifted focus to the middle classes and the elite, at the expense of studying provincial areas and communities.

CHAPTER 2

Sociality between Movement and Space

One afternoon, I sat down with my notebook to try to collect Um Khalid's genealogy. Her two daughters and son sat with us as we brewed some herbal tea on the stove. This was going to be a long session, the son promised, given Um Khalid's age (70), her large family, and the dramas that made their family seem, according to her children, like a '*film Hindī tawīl*' (long Bollywood film). Um Khalid began to list her immediate kin, where they now lived, and who her siblings had married. Her children added details about the characters that began to fill my page. When we got to her fourth brother, Um Khalid could not recall the name of his wife. Her eldest daughter snapped, 'Najwa! That's the name of your sister-in-law! How can someone forget the wife of her brother? You haven't gone senile (*khriftī*), have you?' teasingly suggesting that Um Khalid was becoming forgetful with age. 'Of course I haven't!' responded Um Khalid. 'But when you don't see someone for a time … the proverb makes no mistake: "Far from the eye, far from the heart [out of sight, out of mind]." We don't see them!' After a pause, Um Khalid made a generalized comment about distance, as if to make clear that she was not really blaming her brother. 'Life has changed. *Ma hada la hada!* Nobody is there for anybody anymore. And nobody asks about anyone.' '*A-hoo!* You make it sound like we are city people,' her son replied, partly contesting her exaggeration and partly reassuring her that things might not have changed as much as she feared. 'Arsal is still a village and people are still there for each other. This is what distinguishes the *rīf* (countryside).'

This chapter considers the rural sociality that Um Khalid's son is championing. But how does one study sociality? The variable and diverse use of this concept across the social sciences has cast doubt on its analytical utility (Amit et al. 2015). In the hope of providing a generative framework rather than a fixed definition, Nicholas Long and Henrietta Moore (2012: 41) conceptualize sociality as

> … a dynamic relational matrix within which human subjects are constantly interacting in ways that are coproductive, continually plastic and malleable, and through which they come to know the world they live in and find their purpose and meaning within it.

Christina Torren argues that sociality cannot be an analytical category in its own right because it is a taken-for-granted 'fundamental condition of human being.' Everything we do can be considered as sociality. Yet, how sociality 'evinces itself in personhood and other structuring ideas and practices – kinship, political economy, ritual and so on – remains always to be found out' (2012: 68). The premise of sociality, therefore, is that it is relational, processual, and ethnographically emergent.

Sociality in Arsal is nurtured through everyday moral, affective, and material exchanges. Mundane forms of reciprocity establish the extent to which people are there, or even *exist*, for each other (*al-nās* [mawjūda] *li ba'dha*). The sharing of life through daily interactions builds up what Arab speakers call *'ishra* – the 'bonds of living together' (Abu-Lughod 1986: 63) – a term that translates into 'cohabitation' or 'living with' and subsumes a temporal dimension: living together over time. In describing a similar concept for rural Bosnia, David Henig deploys the term 'consociality'[1] (2012: 14) to emphasize the intimate involvement necessary for living *with* rather than *beside* consociates, those people who 'we grow old with, whose lives we participate in, whom we know intimately and in their own terms. We are entwined with them; we are able to join in their absolutely individual life story' (Carrithers 2008: 167 in Henig 2012). This captures the kind of involvement and mutual engagement Arsalis reference when they take pride in their belonging to a rural society. Endless examples were given to me about the open and trusting nature of Arsalis and 'children of the *rīf*' in contrast with experiences in (or imaginings of) a city like Beirut, where an individual might never see or know their neighbours, let alone engage in social transactions.

But in what ways have these practices changed for Um Khalid and her generation? Through her proverb ('far from the eye, far from the heart'), Um Khalid evokes the relationship between spatio-temporal and affective distance, a relationship that is integral to constructions of social space (Reed-Danahay 2015). Social relationships are inherently spatial. I follow Alberto Corsín Jiménez's argument that social life 'unfold[s] [not] through space but with space ... Space is no longer "out there", but a condition or faculty – a capacity – of social relationships' (2003: 140). An understanding of the physical and spatial transformations of the town over the years sheds light on the entanglements between space, place, time, and sociality. To draw out these connections, I explore different scales in the sections that follow. In the first part of the chapter I trace the ways in which the physical expansion of the town has created capacities

---

1 Henig draws on the work of phenomenologist Alfred Schutz (1967) who distinguishes between contemporaries (those we live next to) and consociates (those we live with).

and conditions for new relationships forged outside of kinship. This was made possible by the decrease in seasonal movement prompted by the decline of agropastoralism, the 1975–1990 war that presented new forms and routes of mobility across the country, and modernization in the town. Following this, I explore everyday post-war sociality through a discussion of the workings of *'ishra* and the resonances of 'being there for each other.' What kind of quotidian social and affective labour is required to honour and maintain sociality? Moving to a different scale, I examine domestic spaces that shape and are shaped by an ethos of 'togetherness.' What people do in their houses, where and with whom, is contingent on seasonality, a growing consumer culture, and gendered ideas about sociability.

1 New Capacities for Sociality

In this section I discuss the expansion of social intimacy beyond the realm of kinship. Kinship has been known to serve as 'an idiom through which to express the power of all social relations considered to have binding qualities' (Bell and Coleman 1999: 6). This is true for Arsal, where kinship provides an ideal model for other relationships and its language is used as a barometer of proximity ('I love my friend so much, she is like a sister to me'). Yet, contextualizing and unpacking the spaces and 'capacities' for social relationships at large enables us to trace social intimacy far beyond circles of kinship. The expansion of the town through space, the movement of residents during the war, and the surge of new employment and educational spaces in post-war years have all led to the widening of spheres of sociality.

Arsal is one of the largest towns in the northern Biqa' region of Lebanon, both in terms of its population (about 32,000 in the early 2000s) and its territory (constituting about 1/22 of the total area of Lebanon [Darwish et al. 2001]). Even today, residents refer to Arsal as *day'a* (village) rather than by the more formal category of *balda* (town).[2] This 'discursive category' (Barlocco 2010: 404) incorporates various – sometimes-contradictory – appropriations that serve as a commentary on conditions of rural life. It is difficult to appreciate the size of Arsal's lands from the town itself, since its residential parts are concentrated in one area, but its lands expand way beyond that. My elderly interlocutors

---

2 The clerk of the municipality explained to me that there is no official differentiation between 'village' and 'town' in Lebanon. What matters is the size of the population, which determines the size of the municipality council. Based on Arsal's population, the town's council is the same size as one of the largest cities of the Biqa', Zahle.

recalled growing up in a time when the physical and social environment of the town was completely different. In the 1940s, I was told, the population was less than 4,000 people and some even claimed knowledge of all the adults who lived in Arsal at that time. They recounted to me how different the rhythms of the town were before (*qabl*), in the old days, when migratory patterns dictated by transhumance shaped people's lives. They spoke of smaller neighbourhoods with very different sociological make-ups and described house interiors that hardly survived at the time of my research.

As the majority of residents practised transhumance, they spent a substantial time away from their village homes, in the highlands and in Syria's lower lands. This made social life feel transient at certain times of the year, as family members moved in and out of the town in accordance with herding cycles. The organization of labour in this agropastoral regime relied heavily on labour units and partnerships made up of kin. It is fair to conclude from conversations with people (and my own observations of a herding family, described in Chapter Four) that, on a daily basis, members of a herding unit were likely to socialize with only a few close kin members for very long periods throughout the year. This does not mean that they saw no visitors, buyers, salesmen, and the rest of it. Agropastoralists are semi-nomadic, which self-evidently means that they mixed with people across the Lebanese/Syrian border and in different villages. But relations of production, intimacy (including marriage), and everyday socializing in general took place predominantly among kin.

Whether among those who returned at certain months of the year or those who resided permanently in the town, sociality was also kin-centred. During my stay, it was necessary to have a car to move from one end of the town to the other. But my interlocutors told me that 'before' people congregated in only two main neighbourhoods, the 'Upper' and the 'Lower,' as they are still called, around the main mosque, which still stands in the centre of the town. Arable fields surrounded the houses, which were all concentrated in a valley encircled by bare, undulating hills. The two main neighbourhoods were constructed around lineages and so were the streets that branched out from them. People still refer to the old neighbourhoods by the name of the lineage that predominantly inhabited them: *hārat* Jawhar, or *hārat* Karama (the Jawhar neighbourhood, etc.). As a family expanded, more rooms, a second floor, and eventually another house would be erected on the same plot of land. In this sense, a neighbourhood could comprise one's siblings, cousins, and second cousins: one's neighbours were also one's kin. When I stayed in Arsal, this practice of expanding on a plot was still popular. It was cheaper to build a new house on land you owned, and parents often wanted their children to stay close to them.

Zuhayr, for example, grew up in his father's old house, which initially had two rooms. His father sold his herds and began to trade goods he bought from Syria. His finances improved so he built another two rooms on the ground floor and, in time, a kitchen and an indoor toilet. When Zuhayr married, his father helped him build a house with two rooms. The building was an extension of his old house. Eventually, as his family grew, Zuhayr built another room. By the time Zuhayr's brother was in his 20s, his father had built a second floor flat on top of his old house with a private entrance, three bedrooms, a large kitchen, and a modern bathroom. Five years later, this was where the brother established himself with his new wife. As his own children began to grow up, Zuhayr laid the concrete for a house on the plot behind his own house for his eldest son. This kind of clustering allows multigenerational families to stay together while recognizing the independence of different households.

As the population increased over five decades, the town started to grow physically. During this period, livelihoods changed substantially and pastoralism lost its primacy, leading to a more sedentary lifestyle. New houses now spread upwards in the direction of the highlands and outwards in the opposite direction towards the neighbouring town of Labweh. Unlike in the old quarters where neighbours (*jīrān*) and kin (*qarāyib*) were often the same, people from different lineages moved in to settle in the new neighbourhoods. Here, neighbourly relations are not based on kinship by default.

The 1975–1990 war played a significant role in transforming Arsali sociality. Political parties tended to consider rural areas as fertile grounds for membership recruitment (See El-Khazen 2003). The appeal of political party engagement had particular resonance in Arsal, as several leftist parties conducted military training in the highlands and had a strong presence in the town. Material signifiers of this era endured in people's homes in the post-war period: Che Guevara posters and black and white photos of Egyptian leader Jamal Abdel Nasser still adorned many a wall. Since political parties operated at a national level, Arsali men moved around the country, sometimes with their families but often as fighters. Political parties provided a forum for meeting people across religious sects and regions in Lebanon. Zuhayr, for example, was one of those who joined the Communist Party. His membership took him to Beirut, where he served as a member of the security team for one of the most prominent figures of the party. Zuhayr frequently told me how enriching this experience was for him. He appreciated that he was in the presence of respected Communist thinkers who expanded his intellectual horizons. One would have never guessed that Zuhayr had only completed primary school, given his eloquence and worldliness. He also felt that the party taught him important life skills: to be tolerant and decent and to be accepting of others. These qualities enabled him to work towards the larger social good of the town in his post-war social activism, in spite of local divisions and tensions. The relationships he forged with other Communists endured after the war. He often received visits from previous comrades and their families who came from Beirut or the south of Lebanon to spend the night.

Within the town, party ties surmounted family and lineage. People who served or fought together developed intense and long-lasting relationships. Zuhayr's comrade Ali, for example, came from a rival lineage. When the war ended, like many others, Ali became disenchanted with politics at large and stopped renewing his Communist Party membership. Thanks to his investment in the stone-quarrying industry, his fortunes changed considerably as he became one of the richer people in the town. Zuhayr, on the other hand, refused to engage in what he considered a dubious and dangerous sector. As an environmentalist, he chose instead to work for a local NGO, despite the low salary. With a family of eight children, he often failed to make ends meet. But Ali always stood by him, offering moral and financial help. He even paid Zuhayr's children's school fees on numerous occasions, with no expectation of repayment. In describing their relationship, Zuhayr felt that his comrade (*rafīq*) and friend (*sadīq* or *sāhib*), despite their now different lifestyles, was much more reliable than his own brothers.

> I cannot depend on my brothers in the same way I could on Ali. My older brother is hopeless. He is hooked on gambling. Can you believe that he has wasted his pension? All that money that the government gave him … wasted. And my younger brother … all he cares about is fashion and he has no job. But Ali, he *yash'ur* (feels) with me. Last Ramadan, he came over with two kilos of meat for our house. He has lent me money so many times … I prefer to spend time with like-minded people rather than my own brothers.

This level of closeness and mutual care, according to Zuhayr, developed as a result of their intimate shared experience, as comrades, outside of the space of the town. Movement out of the town and across Lebanese cities was often described as an educational journey that exposed young men to a life very different from the one they had in Arsal. Zuhayr was keen to visit Communist friends and to maintain his friendship with their families in the hope that the values they had learnt would be adopted by their own children in an age threatened by consumerism and what Zuhayr felt were superficial aspirations.

With the end of the war, Lebanon witnessed a decline in political party membership and a surge of NGOs that mushroomed across the country (Kingston 2013). Political activism in Arsal was replaced by social activism (*'amal ijtimaʿī*) in the newly established environmental and developmental NGOs. These, along with the new occupations created in the town, such as the quarries that provided employment on a large scale, carved out spaces for *zamāla* (colleagueship). In the case of NGOs, this was not just with other town residents but with international NGO employees, other regional NGOs, university professors in Beirut, agricultural experts in ministries in Baʿalbak and Zahle, and so on.

The most important space for forging friendships in the post-war years was, perhaps, that provided by educational institutions. My elderly interlocutors told me that in the past men were taught the Qurʾān in mosques while women received no education. Since then, education has gained significant value. During my research period, it was a common expectation that children of both genders would complete secondary school, at least. As for higher education, the nearest university is in Zahle, an hour and a half's drive away. In the mid-1990s, many university students found it impossible to complete their degrees, mainly because transport was very expensive and students could not pass exams without attending classes on a weekly basis. But in the early 2000s the transport business flourished in Arsal, thanks to the smuggling of diesel across the Lebanese/Syrian border. Many cars, vans, and buses were driven on diesel, making journeys to nearby cities much cheaper. This allowed students to commute on a daily basis to university, which remains a vibrant space for forging friendships.

The spatial transformations in Arsal have created capacities for new relationships. *Zamāla* (colleagueship), *sadāqa* (friendship), and *jīra* (neighbourliness) have become as important to my interlocutors as *qarāba* (kinship). In fact, some of them idealized and even romanticized relationships like friendship, as they were seen to be based on choice and voluntarism as opposed to obligation (Obeid 2010a). Yet it is worth pointing out that these relationships often merged into each other. The professional relationship of *zamāla* incorporated expectations that were had of kin and friends. It was perfectly acceptable for a paid employee at the NGO to leave after a couple of hours of work because her mother was alone at home and needed her daughter that morning. If someone at work was sick, it was expected that colleagues would drop by later to check on them. All these relationships were subject to the rules, expectations, and ethics of *'ishra*, a sociality based on relationality and the 'mutual engagement' (Barlocco 2010: 414) of co-residents.

FIGURE 4   Women cooperating to grind dairy mixture (*kishk*)

2   The Workings of *'Ishra*

Involvement in – 'living *with*' – others' lives requires a frequent, interactive, and reciprocal engagement that lasts over time. I was often reminded of how sharing creates a bond by the (perhaps overused) Arab proverb, 'By living with people for forty days, you become one of them' (*man 'āshar al qawm arba'īn yawm sār minhum*). This suggests that voluntary participation and mutual engagement over time earn one the right of inclusion. But this right requires sustenance. The more intimate people are, the Arsalis contended, the more they 'see' (*y-shūf*; col.) each other; in fact the one is an indicator of the other. Not seeing someone enough is a sign of a loss of intimacy, as Um Khalid's remark, quoted at the start of the chapter, suggested. As an idiom of sociality, 'seeing' entails recurrent visiting, interacting, and doing things with others. The ordinariness of the bonds created through this sharing is reflected in the commensality articulated in the Arab-speaking world as the sharing of *khubz wa milh* (bread and salt). When people share food, a bond is created and, with it, various expectations.

Social activities and reciprocity took place on a daily level. Women baked bread collectively around a shared outdoor stove where they helped each other as they chatted and caught up on each other's news and lives. Neighbours exchanged food; they made sure to leave portions for each other when they made specialties like spinach pies or festival cakes (*ka'k al-Īd*). A man would offer his truck to transport a load of agricultural produce or to move heavy furniture for a neighbour who didn't own a car. These acts of exchange were neither ritualized nor necessarily organized. Rather, they were woven into the rhythm of everyday life as they reassured people that one could still 'knock on a neighbour's door'[3] to rely on a neighbour, and be able to do so spontaneously. The obligation towards neighbours is considered a religious virtue. People often invoked the adage that 'the Prophet has asked us to look after the seventh neighbour' (*al-nabī wassā 'ala sābi' jār*). The number 'seven' is not taken literally, of course. It suggests, rather, a circle of responsibility that is expansive but not exhaustive. The saying is also used to encourage neighbourly relations with nearby villages. One late afternoon during Ramadan, I failed to find a local driver from Beirut and instead took a van that was heading to the Northern Biqa'. As we reached the Baalbak-Hirmil road, the driver felt responsible for me since I was the last passenger and offered to drop me in Arsal, given that it was getting late. But as he began to drive up the hill, he seemed uneasy about entering the town for the first time. He came from a small village in the Hirmil district and had never set foot in Arsal. He had heard rumours that 'the Arsalis were hard people' and was worried that, as a Shi'i, he might not be welcome. He asked me how I was treated and seemed to want reassurance that the people I lived with would not be hostile. I assured him that he would be most welcome in the house I was going to and that he ought to stop for a bite of food and a drink to break his fast. Indeed, when we arrived, Um Yusif's daughter insisted that he parked his van and had something to eat before returning to his village. She assured him that they would never have simply let him drop me and go away, as he was after all a 'neighbour' from the Biqa', and a fasting one for that matter, who deserved their hospitality, for religious reasons if not social.

Other exchanges were more demanding, but people still engaged in them regularly. The intensity of these activities was subject to seasonality, an integral part of Arsali social life. Seasonality consists of 'the movements of people and how the rhythmic structure of their social activities resonates with, and responds to, the periodic transitions of their environment' (Harris 1998: 69). There are two main seasons in the northeast of Lebanon. Winters tend to be sedentary and lethargic. The pace changes considerably in summer, when agricultural activities and household provisioning prevail. During those months,

---

3 The Arabic expression is *du' 'ala jarak*. See Henig (2012) for a similar analysis of sociality through 'locked doors' in rural Bosnia.

I frequently witnessed someone who had dropped by for tea being casually lured into helping out the next day, or over the next few days, to plough his host's land. On such occasions, the return payment was hardly ever more than a meal and some gratitude. As people began to look for market-bought goods rather than homemade ones and adopted advanced farming technologies over customary ones, some traditional activities declined. But until the early 2000s, people still engaged in various forms of cooperation in seasonal and occasional tasks requiring extra labour. These tasks included invitations to shear sheep – still done in 'the old ways,' using hand shears (at one such event I attended, 1000 animals needed shearing) – wash wool, sometimes weighing hundreds of kilograms, in the river to make hand-made woollen mattresses, pick fruit, harvest wheat, and preserve and store food for the winter. These acts of exchange were expressed in the idiom of 'awna (cooperation). The fact that people were invited (ya'zum) for 'awna downplayed the utilitarian approach of inviters seeking extra hands as it simultaneously celebrated sociability. Helpers were offered a meal, often a feast that involved barbecuing a lamb. The idea was that difficult and labour-intensive tasks became lighter *en masse* and that the atmosphere would turn convivial as people joked, sang, talked, and snacked while working hard. My participation in these activities showed me that, for the Arsalis, there is virtue in *jam'a*; a gathering or the ability to bring people together is in itself rewarding, whether in the context of agricultural cooperation or social events.

While forged through everyday engagement and reciprocity, sociality was epitomized in important life events such as marriage and death. Kin, neighbours, friends, and colleagues participated in these important junctures, for, as people often told me, it was in happiness and sadness ('the bitter and the sweet') that co-residents demonstrated mutual engagement. Weddings are a good example, for they were one of the most inclusive events in the town, often held around the second half of August, marking the end of the agricultural season when cash was readily available. Families sent out verbal invitations, but these were usually addressed to whole households, 'tfaddalū 'ala 'urs ib-nina' ('We are inviting you to our son's wedding'). In other words, it was difficult to anticipate exactly how many guests would be joining the celebrations over a three-day period. The parents of the bride and the groom, each in their own house, were expected to cook a meal large enough to feed however many people might turn up. For three days before her daughter's wedding, Um Naji's kitchen was full of women of various ages from the neighbourhood, kin and non-kin. Along with her sisters and daughters-in-law, they spread out over the kitchen and two more rooms preparing the wedding feast, some rolling dolma leaves, others making stuffed *kubba* balls (lamb croquettes). Outside, the men erected a marquee, connected loud speakers, and borrowed trucks to transport chairs for the party. Um Naji was not from the predominant lineage of her neighbourhood. Yet she was very moved by the efforts her neighbours took to make her daughter's wedding a success. Her neighbour's son, a close friend of her own, promised that he would personally make sure the wedding was a celebration to be proud of. He and his friends blocked off the street to make a dance floor large enough to contain the line of a hundred men and women holding hands, shoulder to shoulder, as they performed the *dabka* (a folk group dance).

It was precisely the extent of all this effort that brought home the amount of moral obligation within neighbourhoods to respect the sanctity of death, for weddings were immediately cancelled as soon as the call to prayer announced someone's death. The music would suddenly cease and the guests would disperse upon receiving the news. For families who might have spent their year's savings on a wedding, this was a devastating blow, but respecting the grief of neighbours was always a more pressing concern, even when the bereaved were not close. The spatial development of the town into dense neighbourhoods meant that the noise from a party could spread over several neighbourhoods, and the likelihood of having to cancel a wedding party always loomed large.

As soon as the news of a death was out, crowds of people immediately went to the deceased's house, even before the burial. Men and women mourned separately for the first few days. The rooms of the house would be filled with people sitting in silence. In the women's section, once in a while someone, usually an elderly woman, would wail and sing the *'atāba* (a song of loss or longing). Since the deceased's family was usually in shock, it was kin and neighbours who cooked or brought with them large pots of food and bread. The idea was to show the family that their grief was being shared. At funerals, men carried the coffin to the cemetery and helped to bury the body; an act that the Arsalis believed earns them *ajr* (religious recognition).

Failure to participate in celebrating a wedding or showing grief at a funeral could cause offence. The latter especially may be considered a 'betrayal' of *'ishra*. This is exactly what happened when Zuhayr died. His family and friends were devastated by his unexpected death. The NGO that employed him shut down for three days as a sign of mourning. Many friends and co-workers came to the funeral from all over Lebanon, a reflection of Zuhayr's relationships built up through years of political and social activism. Yet it was noted that one of the experts Zuhayr had worked with for over ten years failed to show up at the funeral or, it seemed, to have offered his condolences to either Zuhayr's family or his colleagues at the NGO. This created a deep wound, as *'ishra* was seen to be wasted (*di'ān al 'ishra!*).

It is not my aim to paint a romantic picture of rural life by presenting a 'sentimentalised view of sociality' (Edwards and Strathern 2000: 152) or downplaying social conflict in Arsal. But I do want to convey the shared concern that people felt for ensuring the continuation of harmonious living so as to enhance 'being there for each other.' In this sense, much effort went into avoiding, resolving, and downplaying conflict (c.f. Gilsenan 1976; 1996). *'Ishra* calls for an appreciation of time, the value of which adds to the process of *living with* other people. I suggest that this temporal investment is cherished in its own right for turning *'ishra* into a buffer, or at least a rhetorical device, against conflict. This became clear to me when Um Yusif's daughter once quarrelled with a neighbour who had taken her daughter's jacket, claiming she had mistaken it for her own child's. The daughter accused the neighbour of stealing and, as the quarrel escalated, asked her to leave. When she found out about it, Um Yusif was disapproving of her daughter's behaviour, regardless of who had been at fault. Invoking the supremacy of *'ishra*, which demands that neighbours rise above petty conflicts, she told her daughter to make some pies and take them to the neighbour as a sign of reconciliation. For Um Yusif, neighbourliness through *'ishra* was by all means more valuable than the jacket, or an argument over its ownership.

Reciprocal engagement remains highly valued in the town as it gives meaning to everyday life. *Living with* others, 'seeing' them regularly, and being mutually involved in a web of social and communal bonds is seen to be the quintessence of good rural living.

3   Domestic Spaces, Gender and Consumption

If proximity is enacted by 'being there for each other,' the domestic is an important space that shapes and is shaped by this ethos of togetherness. Daily practices, actions, and movements within a house can tell us a lot about social relations and the place of gender, age, and local cosmologies in spatial organization (Bourdieu 1977). These are enabled by the materiality of a house, its architecture, and interior decor, all of which were changing in Arsal in the years following the war. With a growing consumer culture, the household has become a key arena in which a distinct form of modernity is generated (Lewinson 2006). Not only were people investing in embellishing their homes, but the way they lived together 'today' was a matter of stark contrast with the 'old days.'

3.1   *House Interiors*

My elderly interlocutors recounted that crowded living used to be the norm in the old mud houses. Um Khalid, for example, told me that, at certain times of the year, when the herders returned temporarily to the town, her parents, four brothers with their wives and children, and four sisters, shared the house, sleeping in one square room. The toilet was a tiny room containing a hole in the ground, built outside the house at the end of the front yard. The house had no kitchen. Cooking was done collectively on the fireplace in winter and outside the front door in summer. Um Khalid associated the provision of space today with a better life and better relationships. Talking about her childhood, she told me that her older sister had been cruel to her younger siblings. 'She caught me with my younger sister stealing jam. They had instructed us not to touch the jars. My older sister was so angry with us, she caught us and rubbed chillies on our … [she pointed to her crotch]. *Yī*, our screams! What can you do? We lived on top of each other in those days.' Um Khalid related her sister's brutality to the physical space itself. In hindsight, she felt that the cramped space must have had a toll on how family members dealt with each other. This contrasted very markedly with how tolerant members of her own household were towards her grandchildren, four of whom lived in the same extended house. Reflecting on how different social practices were in her youth, she pointed out that there was no *khusūsiyya* (privacy). In her humorous manner, she described their sleeping arrangement as *rās wa tīz* ('head by ass'). 'The married ones, if they had some business to attend to [have sex], the rest took no notice or they just found a place, like in the closet' – a small room that contained grain storage tubes called *khalāya* (cells). In our conversation, Um Khalid established that living standards were far better today, whether due to having more space or more comfortable amenities. Yet she was critical of what she thought was an excessive consumerist attitude towards house interiors. 'It is not that people didn't have taste (*dhawq*) back then. We cleaned everyday and tidied up. But things were more *'al-tabīʿa* (natural, spontaneous). We all lived in mud houses and, if anything, you would nail some ornament on the wood. Nothing like today. People spend months deciding on just the colour of their curtains!' Very few old mud houses remained at the time of my fieldwork. Their occupants had become used to visits that showcased to outsiders 'the old house' as a remnant of the town's living heritage (*turāth*).

By the 1990s, domestic architecture had changed significantly. Houses were built out of concrete instead of mud, with the development of the stone quarries making stonework exteriors a desirable aesthetic. Although very few people could afford to complete a house by the time a newlywed couple moved in, ideas about living space had changed significantly from the conditions Um Khalid described. A minimum level of habitability, and new ideas about 'essentials,' necessitated having at least one furnished room, an internal bathroom with a pedestal toilet instead of a squat toilet, and a kitchen with modern appliances. Even when a couple lived in an extension of the parental house,

FIGURE 5   An old mud-house: 'remnant of living heritage'

building a separate kitchen marked the independence of the new household and, in particular, the domesticity of the wife. While these families might still cook or eat together, a wife is expected to take charge of cooking for her own family.

Much effort was put into the choice of interior furnishings, as Um Khalid commented. Patterned curtains matched the colours and motifs of floor cushions. Paintings of birds, waterfalls, and the Ka'ba in Mecca decorated people's walls. Materials and textiles, and the fashions determining them, were imported from Syria, especially Homs, a major destination for shopping. Before the improvement of transport in the Northern Biqa' region, very few shops existed in the town in the early post-war years. The ones that did were often an extension of somebody's home and stocked the few items that were considered essential (tea, sugar, coffee, soap, matches, candles, and some fresh vegetables). The nearest markets to Arsal on the Lebanese side were in the city of Baalbak, but residents found Syrian markets to be cheaper and observed that the same but pricier merchandise was often found in Biqa'i shops. It was more sensible to shop in, or rather from, Syria. In the 1990s, the neighbourhoods I was familiar with relied on a private delivery system run by one man, Said, who worked as a bus conductor. Instead of receiving a wage, he struck a deal with the driver to use his vehicle to transport goods. He compiled shopping lists from the different households in a neighbourhood, returning at the end of the next day with goods ranging from items of food (industrially produced bread was particularly desirable, as were Syrian specialty desserts) to pipes for wood burners. Said added his fees to the bill and, after a few years of pursuing this business, did well enough not only to take a second wife, but also to go on honeymoon for a week in Syria – something that led to accusations of his having become '*mutamddin*' (acting urban, or modern) since this practice was popular in other parts of Lebanon, but not among Arsalis. For more substantial shopping, women took day trips together to Homs to look for bridal jewellery, textiles, material for floor divans, trousseaus, and bridal dresses. People were able to recognize the shops in Homs from which particular materials and designs had been bought. They compared notes and commented on what was and was not *dārij* (in fashion).

Through observing spatiality within a house, we can trace how the use of domestic space transforms social interactions (Pader 1993; Mughal 2015). While the aesthetics and uses made of domestic space seemed to differ from those described by the older generation, the modernity of these new houses did not have as their purpose the separating out of activities in the way we see in modern urban design – rather, the practices I observed in domestic spaces reflected a social context in which ideas of sharing and togetherness still held sway. Most of the houses I visited had a similar interior design. Houses had several rooms, furnished almost identically, serving multiple purposes: cooking, eating, sleeping, studying, relaxing, and entertaining were all done in the same space.

The three rooms in Um Yusif's house were similar in design, but with different colour schemes for the furnishings. 'Arabic' divan corner seats sat on a thick hand-made carpet with traditional Arsali motifs that covered the entire floor space. An alcove called a '*yok*' (col.) occupied the centre of the main wall. It was used to stack the thick hand-made woollen mattresses that were an essential component of bridal trousseaus, and was covered with patterned curtains matching the divans. In all three rooms, two floor-to-ceiling cupboards flanked the *yok*. They stored clothes, sheets, towels, and the personal belongings of the different family members. The mattresses were pulled out at night and folded and put away in the morning, thus transforming the room from a sleeping space back to a sitting room. In other households, sons and daughters slept separately, especially when they reached their teens. Ideas of 'privacy' may have changed from Um Khalid's day, but sharing space was still the driving ethos behind living together. No family member, not even children, owned a single room. All the spaces in the house were for everybody to use. And, as

I point out in more detail in Chapter Six, the desire to be alone in a room or a space, away from the group, worried other family members.

By the early 2000s, some of my interlocutors had begun to alter their furnishings slightly. Some added a sofa to the divan. Others began to buy beds. For example, Siham had decided that she wanted a European bedroom (*'ūda franji*). But when I stayed with her one night, she offered me her queen-sized bed, which she said had hardly been used. She, her husband, and children carried on sleeping in one room. Siham commented jokily that '*al-moderne*' did not really suit them and that they all preferred 'the old way,' sleeping on the warm floor mattresses and spending time together in one main room. It was usual for the women to prepare food in such rooms, especially in winter when families huddled around the *sūba* (wood burner) to save the expense of heating the whole house. The kitchen was used for quick tasks or for frying food to avoid strong odours in the main room. I was told that the quality of food is always better when cooked on the stove of the *sūba*, and that a room does not really warm up until the aroma of a *tabkha* (a stew) fills the air.

When the food was ready, the plates were placed on a plastic mat spread on top of the carpet in the middle of the room. The people sitting around the mat were each handed a loaf of flat bread and, unless the food was sloppy, they ate out of shared plates. At the beginning of my research, some of my hosts teased me about having to eat on the floor, rather than at a table, and asked whether I agreed that food tasted better from a shared plate. Once someone had finished their meal, they withdrew to one of the divan seats and carried on with the evening. A typical scene on a winter's evening at Um Yusif's house saw a couple of family members watching television, while the older children did their homework or played with their younger cousins, and neighbours or relatives dropped by for tea. I often found it amusing (and sometimes confusing) to follow the conversations taking place simultaneously: one would be discussing diesel prices, another describing the happenings in the latest episode of a dubbed Turkish soap opera. The children were yelled at when they began to misbehave. The noise mingled with the smells of cigarettes, burnt diesel, and the endless stream of delicious snacks and meals that kept appearing from different sides of the stove. The world of the adults was not separated from that of the children, who took part in the conversations about breaking news or contributed details they had heard on other occasions. Once, after the evening news, I heard the adults listening seriously as Um Yusif's nine-year-old grandson began to argue that Muhammad Said Sahhaf, Iraq's Minister of Information, was exaggerating the claim that the Iraqis were beating the Americans. Perhaps echoing views he had heard elsewhere, he entered into a heated debate with his grandmother, who maintained that God would ultimately give victory to the Iraqis, in spite of contradictory news stories. That evening, his six-year-old sister was more concerned with devil-worshippers. She had heard a story on the news about a group of youths arrested in the capital for listening to hard rock music. She demanded to know where exactly the devil lived!

In spite of sometimes expressing relief when these noisy evenings ended, most people regarded the mingling of multiple generations, neighbours, and friends in a single room as a sign of a healthy social life that celebrated togetherness and the *jam'a* (gathering).

3.2   Gendered Sociality

The *jam'a* described above took place on a winter's day. The changing rhythms of the seasons determined when, where, and with whom men, women, and children socialized. The *sūbas* had been removed by mid-April as the weather warmed, changing the 'involuted sociality' (Harris 1998: 76) of the winter room. The pace of life in the town became faster and more energetic. The streets became busier and filled with pedestrians, cars, motorbikes, and trucks travelling to and from the highlands. Young men loitered in the main square, children played football on the streets, and elderly men sat in front of their houses and watched life go by.

Within the house, family members still gathered at different times of the day, especially at meal times, but socializing seemed to spread out spatially and across lines of gender and age. Now that different spaces in the house could be used, men and women socialized separately, though this was not obligatory. Inter-gender friendships were accepted in Arsal. Many students formed study groups and met either in the local youth club or at somebody's house. Male friends visited their female counterparts, single or married, as long as they sat with the family rather than apart. But on a warm night it was not uncommon to see *shabāb* (young men) on the balcony of a house enjoying an *argīla* (water pipe or hookah) till after midnight. In one of the rooms, a few women might entertain other women of the same age. They would watch a television programme or chat and snack. These spaces were not segregated. Often a room would fill up with people of different ages. As the evening proceeded, a son or daughter might invite his or her visitors to move elsewhere. 'Leave the *khatāyra* (elderly) to their own stories,' younger people would say as they relocated in the next room. Or, 'Let us women move in order to feel more comfortable.' These manoeuvres were carried out delicately, without causing offence. For example, a daughter would make sure that her father's guest had been offered the appropriate hospitality before moving into the next room with her own friend. The move itself was acceptable because people appreciated that men and women liked to socialize differently.

The common assumption was that men and women were interested in different topics of conversation. Men talked too much about politics (*siyāsa*), women would complain. Men presumed that women's interests were domestic, which didn't always interest them, or they felt that women's conversations were limited to the scale of the neighbourhood and were too gossipy – a dubious story might sometimes be derogatorily dismissed as *sawālif niswān* (women's talk). These gendered stereotypes did not necessarily reflect the breadth of interests and range of either men's or women's conversations. Nor did they represent the varied cross-gendered conversations that took place in reality on an everyday level. But it is worth taking seriously these gendered ideas about sociability and the corresponding practices that reproduce and convert certain relationships. In proposing sociability as a domain for analytical inquiry, Sally Anderson (2015) suggests that sociability is 'infused with form' and in order to investigate this concept in an anthropologically productive manner, she advises that we 'probe all forms ... as conceived and practiced by the people with whom we engage'.

In Arsal, one such form that throws light on how gendered spaces and practices are constructed is the consumption of particular drinks when socializing. Regardless of the time of day, when a visitor drops by in winter, the host usually offers them a hot drink of tea, coffee, or *mate* (also known as *yerba mate*), the South American herbal drink that is very popular in Syria and Lebanon. A cold juice or carbonated drink is offered first in summer, and a plate of mixed nuts and seeds and some biscuits are usually served as an accompaniment to the drink. After this, guests are continuously offered food and drink for as long as they stay. In the course of the long afternoons I spent with the families I was familiar with, I would be offered seasonal fruits, a loaf of freshly baked bread, more drinks, fresh cucumbers brought from the field, a main meal, another round of drinks, and so on. These offerings are referred to as *tislāya* (leisure, or passing time pleasantly), which is how a host will urge a guest to carry on helping themselves: '*Tsallī, tsallī!*' I would be pressed as soon as my hosts noticed that I had allowed a second to lapse without touching the plates of snacks. By continuing to consume what is on offer, the guest takes part in sharing the leisure and pleasure of socializing, and thus shows appreciation of the host's hospitality. This exchange applies across the board, with both genders and all ages. Of the hot drinks, *mate* was the most popular in the town, and everybody I knew consumed it. Yet men claimed that *mate* was a women's drink and that they preferred coffee or tea. The rationale for this had to do with local ideas about consumption and time and how the different genders enjoyed, used, or 'wasted' it.

Coffee and tea, which are brewed and offered in individual cups, can be consumed quite quickly. Men visiting on an errand, for example, were able to accept a cup of coffee as a sign of hospitality since they could leave soon after. An offer of *mate*, on the other hand, lends itself to a longer social session because of its method of preparation and consumption.[4] *Mate* breaks are taken several times a day, whether among members of a household or when visitors are present. The host takes charge of the '*mate* tools:' a tray with a kettle full of hot water just below boiling point, a bowl of loose *mate* tea, a bowl of sugar, the *mate* cup or gourd, and a stainless steel straw. Half a cupful of tea is steeped in water with one spoonful of sugar. The host usually sips the first cup through the straw before adding half a spoonful of tea, half a spoonful of sugar, and more hot water with each serving. To cleanse the straw, some hosts rub it with a slice of lemon or pour hot water on it before passing it on to the next person. This process is repeated and only ceases when the drinker indicates they have had their fill by shaking the cup twice and exclaiming '*graysa!*' (a version of the Spanish 'gracias!'). If someone new drops in (male or female), they take their place in the round. The more drinkers there are, the longer the session takes as the cup circulates around the room. Of course, it is possible for someone to drink only a single cupful, and the women often did if they were in a rush to get on with their work. The issue here is not with what is being consumed, but rather its manner and frequency. The perception is that by default a *mate* session takes up time through its potential for leisurely waiting, during which women gossip, chat, and relax as the cup makes its way towards them. The drink, I suggest, serves as a way of understanding gendered perceptions and attitudes about how time (and spending it) has changed for men and women.

This was made clear to me one day at the NGO. One of the senior male employees declared that *mate* should be banned in the town as he held the drink responsible for the loss of female productivity. The time women spend on *mate*, he argued, could be used in more productive endeavours (*intāj*). Given that the women he was addressing were already employed as artisanal workers in the NGO as well as running their households, I invited him to expand on what else he thought women should be doing. Admitting that he was probably referring to 'other women' and that the local job market was tight for women in general, he reminisced on the productivity of their foremothers and how women herders 'never stopped for a minute.' Instead of buying things from

---

4   In Lebanon and Syria, different groups have different *mate* preparation rituals; for example, some might serve the drink in individual cups. In Arsal, the cup is shared.

FIGURE 6   A woman preparing *mate*

the shops and wasting time on gossip, he reminded us, they made jams, yogurt, and cheese. One of the women listening to him seemed fed up with his moralizing. She teasingly advised him to calm down and join them for a round of *mate*. This broke the tension and everyone laughed as he sheepishly conceded that it was indeed time for his cup of *mate*.

Although provocative and dismissive of women's contributions to social life, the employee's comments encapsulate some significant interrelated transformations that have taken place with the decline of agropastoralism over the years. The first has to do with the amount of time that has been freed up for women now that they are no longer required to perform the intensive, year-round labour that constituted an essential part of their domestic duties. The time now to be enjoyed in an emerging model of domesticity (Chapter Four) is made possible by their entry into a growing and accessible consumer culture that has transformed household tasks with the acquisition of such articles as washing machines, cookers, and refrigerators. The improvement of transport in the North Biqa' region, which has both increased cross-border shopping and led to the opening of new shops in the town, means that items previously made in the home, like mattresses, carpets, and food, can now be bought in the market. Consumer goods thus became a signifying practice that has transformed gendered identities (Forte 2001; De Grazia 1996; Miller 1997, 1998) and allowed women more, and more frequent, leisure time. The issues raised by the male NGO employee point to a wider ambivalence about the relationship between shifts in consumption and production, and sociality. My interlocutors debated whether, beyond considerations of practicality and signification, *khālis* (readymade) products were better than homemade ones. They wondered if homemakers who still laboured to bake bread and make jams invested more in their families. This was particularly obvious in discussions about food since most food provisioning (*mūna*) was done collectively during sessions involving *ʿawna*. Women like Um Khalid feared and worried that, after they were dead, their daughters would never again enjoy homemade preserves. At an analytical level, distinguishing between consumption and production is perhaps futile (Graeber 2011). At an ethnographic level, however, these distinctions were used to comment on palpable shifts in the town and in gendered relationships, particularly within the household. With the withdrawal of women from older production regimes, my interlocutors, of both genders and different ages, were reflecting on their roles as men and women in their own social worlds. These discussions were not just about the quality and durability of things; they were moral questions about practices that people like Um Khalid saw as corroding sociality. This, perhaps, was the sentiment she was conveying when justifying why she had forgotten her sister-in-law's name, as described at the beginning of the chapter. In her experience, the nature of being and doing things together – certain notions of 'living well' – had changed. In the next chapter, I look more closely at the changing nature of livelihoods, and how these have shaped what it means to live well.

CHAPTER 3

Living Well: Experiments in Livelihoods

It was a quiet day around noon towards the end of June 2003, nothing unusual for a time of year when people hide indoors from the heat of the sun. What I did find unusual was finding Salim and Kamal sitting in a tiny spot of shade on the porch of the NGO. They were still in their dusty work clothes, smoking and sipping tea, perhaps counting on the combination of caffeine and nicotine to appease their anger over the events that had sent them home that day. Both men worked at one of the quarries that had sprung up on the fringes of the town since the end of the 1980s. For months many rumours had been circulating about the fate of the quarrying industry. The government had been attempting to issue a land-use law in order to regulate this sector. But the law, so far, had not been finalized, though it had decided that all activities must cease until further notice in an attempt to prevent further abuse of communal land in the Arsali highlands. Now matters had taken a serious turn – checkpoints had been set up at different locations in the town and labourers sent back home. There was a feeling that a conspiracy lay behind the vagueness surrounding the legislation – or so Salim believed.

> We all know who will benefit from these laws. You think the Arsalis will? The Arsalis will be made slaves on their own land! The big industries that can afford to pay 40,000 USD will get the licence and employ us as manual labourers. But this is something that we as Arsalis will not allow. We should think of ways to object!

When I tried to probe deeper into the nature of these laws, or fathom where the exact amount of 40,000 USD had come from, I found I was not the only one at a loss. I learnt that the Ministry of the Environment had indeed drafted a decree (no. 8803/2002) in 2002 to regulate policies relating to quarrying activities, but although the Council for Development and Reconstruction developed a National Master Plan (*Mukhattat Tawjīhī* or *Schéma Directeur d'Aménagement du Territoire Libanais*) in 2004, it was not approved by the Council of Ministers until June 2009 (Darwish et al. 2011). By then, the quarrying industry across Lebanon had expanded without regulation.[1] Information the town received in the meantime was based more on hearsay than actual decisions. Thus, when

---

1 The number of quarries in Lebanon doubled between 1989 and 2005 from 784 to 1278 – i.e. from 1 to 1.8 quarries per 14 square kilometer (Darwish et al. 2011: 351).

Salim and Kamal were sent home because of an uncertain law, it played on post-war anxieties about livelihoods and confirmed their sense of marginalization. 'What I wish to understand,' continued Salim passionately, 'is what kind of planning this state has, stopping people's livelihoods like that without offering any substitute? For us, it is obvious that it is enforcing *siyasat tajwīʿ* [a hunger policy].' Overwhelmed by mixed feelings of disappointment, anger, and uncertainty, the two men, like many others in their position that day, racked their brains for solutions. 'It's time for us to think of what we can do now. We have to think of a big project that will benefit all Arsalis. We need a *badīl* (an alternative). We've tried cherries, it hasn't worked, we've tried smuggling, it hasn't worked, and now quarrying doesn't seem to be working.' Kamal, on the other hand, pessimistically concluded that 'The Arsalis are doomed!'

Following the decline of agropastoralism in the 1970s, the Arsalis began to diversify their livelihoods by 'construct[ing] a diverse portfolio of activities and social support capabilities in order to improve their standard of living' (Ellis 2000a: 232). Such diversification may take place for a variety of reasons, sometimes voluntary, in order to increase income, and sometimes out of necessity, as a result of economic, environmental, or political factors. It is often the case that a number of overlapping processes, such as seasonality, changing labour markets, and migration, will lead to diversification. Rural livelihoods have been known to change rapidly over time (Murray 2001). To understand these changes, we need to consider the intersections of 'the micro-level of the household, the meso-level of institutional intervention through local government, development agencies or regional markets, and the macro-level of national policy-making' (Murray 2001: 4; Scoones 2009). In the Arsali context, changes in livelihoods were mediated by a multiplicity of forces: the 1975–1990 war, reverse migration towards the end of the war, the post-war state agricultural policies (or lack of them), the political situation in the Northern Biqaʿ by which people lived under the actual rule of two states (Lebanon and Syria), the aridity of the land, population increase, and the inevitable pressures of modernization.

Most households I knew opted for a multiple livelihood strategy. When describing their livelihoods, my interlocutors often used the vocabulary of experimentation (*tajruba*; *tajārub* pl.) – the 'trying out' of new avenues such as the cherry farming that Salim mentioned. As they saw it, their *tajārub* were driven by chance, politics, and fatalism. As I will show in this chapter, the *tajārub* approach reflects at once the pervasive vulnerability that is intrinsic to rural livelihoods (Ellis 2000b: 298) and a sense of innovation and relentlessness acquired in the face of adversity and the need for survival. '*Al-rizq ʿala Allah*' ('Leave the livelihoods to God'), people would repeat, a saying that may give the impression of passivity or docility. But the saying is countered, often

FIGURE 7   Fruit orchards in the highlands

emphatically, by its second half, '*lākin al-ʿabd lāzim yasʿā*' ('but the believer must persevere'). It is this contradiction – between acceptance of the harsh conditions of their livelihoods that bring *taʿab wa shaqāʾ* (tiredness and weariness) and their ability to endure hardship and create new opportunities – that defines how people construct themselves as social agents.

A relatively successful venture, quarrying had proved to be a stable and important source of employment for Arsali residents in the 1990s. The fear aroused by the dubious law led to a contagion of pessimism as people felt that yet another experiment might have failed. Salim's remarks about the state's intending to impoverish locals ('hunger policy') harked back to the town's historically tense relationship with the state.

In the first part of the chapter I will outline the history of production that led to the diversification of livelihoods and the rise of quarrying in and around Arsal and explore how the anxiety generated in the town by the law sheds light on state-society relations. The second part of the chapter examines the moral economy of the new livelihoods. The shift away from agropastoralism has had repercussions on family relations, household organization, and people's relationship with nature. What has been the impact of moving away from livelihoods built on collective and collaborative labour that capitalize on the commitment of family members to ones that are the remit of individual

LIVING WELL: EXPERIMENTS IN LIVELIHOODS 43

FIGURE 8   A family picking cherries

males? The chapter addresses the conflicting responses and values attached to these diverse livelihoods.

1   Livelihoods as an Ongoing Experiment

In the recollections of local people, there was a specific serendipitous turning point in their history of production. In the late 1960s, at a time when the major livelihood was agropastoralism, one man took the somewhat haphazard decision to plant cherry trees in the highlands as an experiment, a venture viewed with scorn in the early days. 'They mocked me at first, they invited each other to see the crazy man (*al-majnūn*),' Ali, who has since died, told me. He continued proudly, 'Then they all followed suit because I proved to them that it is lucrative and the cherries … nothing more beautiful!' By the mid-1970s, people had started to embrace the idea of cherry farming, without necessarily abandoning herding, though as Um Khidr explained, pointing to her orchard, some did so gradually:

> We used to plant this orchard with barley, wheat, and other grains. When our neighbours started planting cherries, we felt encouraged but were

> afraid to plant all our fields with cherries at once. So we decided to split our land and planted some cherries here and some grains there. After six years when it proved profitable, we planted some more cherries and hired a shepherd. Then, some years later, we sold the herd. Since then, we only come to the orchard and stay in our *jurd* (highland) house during the cherry season.

Since the 1960s, an estimated two million fruit trees have been planted in the highlands of Arsal (Talhouk et al. 1996). At its outset, fruit production inspired new confidence in the Arsalis, giving life to hopes that started with the reforestation and improvement of their dry land and led to the promise of an entire new lifestyle. When I asked Um Khidr whether she preferred fruit growing or herding, her reply was quite decisive.

> The cherries are much better because they are less tiring. You can work at your own pace in the shade of the trees. If you're tired, you stop and take a nap, you have a cup of tea with whoever is helping you. There is a great difference between the two.

Fruit growing relieved people of the intensive labour and long working hours that herding dictates. By the 1990s, with the increased use of technologies, cherry production took up no more than three months of labour per year (Darwish et al. 2001).[2] By the end of the decade, agriculture ranked above pastoralism in economic importance, providing for some 3,375 households in the town (Hamadeh et al. 1999). The seasonal nature of cherry production let people pursue new opportunities and diversify their livelihoods.[3] Although these opportunities promised less intensive labour and relatively easy profits, some of them brought with them risk and loss. Two major livelihoods, smuggling and quarrying, flourished throughout the years of the 1975–1990 war. The first was made possible by a border that was barely monitored, and the second by a surge in the expropriation of communal lands, carried out with impunity.

During the 1980s, a time when the local authorities were unstable, smuggling intensified across the Syrian–Lebanese border. Products ranging from tobacco to washing machines were brought in illegally on donkeys, tractors,

---

2 By the mid-1990s, fruit production also took over grain production, which was most popular in the 1960s. Darwish et al.'s survey, which studied land distribution between trees and traditional crops, shows that 78% of cultivated land is allocated for tree crops whereas only 22% is allocated to traditional crops (i.e. grains) (2001: 96).

3 By the end of the 1990s, less than 35 per cent of farmers earned their income from agriculture alone and 60 per cent had at least one other form of livelihood (Hamadeh et al. 1999).

and pick-up trucks. Even children were involved: 'I used to stack Kleenex boxes on our donkey and cross the border,' one interlocutor told me as he recounted his childhood adventures. By the 1990s, the Syrian state had tightened its rules. After the Ta'if Agreement brought an end to the war, Syria pledged to support Lebanon by ensuring that no arms were smuggled across the border. Although smuggling still took place, it had become a high-risk enterprise. In the summer of 1997, I visited an area in the *jurd* (highlands) that was a small hub for smuggling. Only six houses stood in the vicinity, about a kilometre or so apart. My friend's mother was not a smuggler herself; women did not engage in this dangerous activity. But that year she was raising a herd of 20 sheep – too many to graze around her house in town, so she decided to use the family's *jurd* house during the summer months, after which she would return to Arsal to sell the animals. Her daughter invited me for a week's visit to keep her mother company. During this time, we visited our neighbours – mostly men who were spending a few days here on a particular operation before returning to their families in town. One man and his wife lived permanently in the area as they had eloped and were not yet reconciled with their families, preferring to keep their distance. On one visit, we were welcomed by a man who invited us into the living room, which turned out to be filled with merchandise. The whole house looked like a shop, apart from the few rifles standing near the door – '*lawāzim al tijāra*' (tools of the trade), he explained. Two Syrian men had just arrived on their motorcycles from Qara, a town on the other side of the border. He described them as his trading partners. On one of the nights of our visit, we were awoken by the terrifying sounds of gunshots and screaming car engines. 'Here they go with the *hajjāna*[4] (border police)!' commented my friend's mother who was obviously familiar with the noise of a police chase. By the time we woke up in the morning, she had already had news that no one had been hurt the night before.

When I returned to Arsal again in 2003, this small community of smugglers had left the *jurd* and returned to the town. It turned out that dozens of young smugglers had been shot, captured, and incarcerated in Syrian prisons. Some, however, still pursued it in spite of the risks, for the business of smuggling carried its own attractions. Mazin, a long-time smuggler, used the analogy of cigarettes to express the smuggling 'addiction:' 'I only have to work for a couple of hours and make money enough for a week.' He said that during the good days of the late 1980s he could make up to 25,000 USD per operation. 'It is dangerous,'

---

4 This is not the official term for the Syrian border police, but rather the one used by residents on both sides of the border. The word itself means 'camel riders' and may have some historical reference.

he went on, 'I could get shot, but it is like smoking. You know it could kill you, but you still do it.' By 2002, however, having 'faced death' in his last operation, Mazin had sworn to give up smuggling forever. 'I was so close to being caught in Syria the last time ... You could be imprisoned or [shot] dead. *Ma bi-yistāhal!* (It is not worth it!). From that moment, I swore I would stop forever.' Those wise enough to quit smuggling decided to invest all that 'easy money' in profitable projects. Mazin himself built a petrol station, though others tended to put their money in the burgeoning quarrying industry. I knew several people who had worked as smugglers but missed their opportunity to invest. Some had spent their money on building big houses, still unfinished during my stay, and found themselves lamenting their short-sightedness. 'Naïve (*sāthij*) people like me thought that smuggling would last forever with the same profits. But things have changed and it is no longer lucrative,' a former smuggler admitted to me.

The other growing employment opportunity, quarrying, would have been unthought-of at the time when fruit farming seemed so promising. But this industry was brought to life by the process of reverse migration that took place at the beginning of the war. Many Arsalis who had moved to the suburbs of the capital before the war, sought refuge from the fighting in their hometown, which escaped the violent battles taking place in Beirut and other parts of Lebanon. It was the new ideas and experiences they brought with them to the town that prompted the start of quarrying (Baalbaki 1997). When I interviewed the municipality clerk (*Kātib*) in 2003, he told me that this new sector accounted for a significant number of labourers, about 4,000, if we considered all the related enterprises – sand processing plants, transport (namely trucking), and associated crafts and services such as stonemasonry, electrical and technical maintenance, etc. The financing of quarrying came from the shrinking smuggling industry.

By the mid-2000s, most Arsali households had adopted a multiple livelihood strategy – even herding households were economically diversified. And most, about 75%, of the Arsali labour force still worked in agriculture (Darwish et al. 2001). But diversification is not a strategy that is temporary or transient. The vulnerable nature of rural livelihoods meant that diversification had become a necessity. For example, fruit farming[5] was not lucrative on its own. The climate frequently led to crop failure, with snow falling right up until April in some years. In addition, harvesting time was delayed because the fruit orchards were generally planted at very high altitudes, and this, together with the bad condition of the roads leading to the highlands, made marketing a

---

5 Fruit farming included cherries, which occupied about 36% of the cultivated areas of Arsal, apricots 26%, grapes 14%, and wild pears and olives 24% (Darwish et al. 2001: 96).

challenge. Farmers were compelled to sell at low prices once their crops were ready through middlemen who rarely gave them a fair deal (Darwish et al. 2001; Hamadeh et al. 1999). The situation was exacerbated by the lack of policy-making and regulation in the agricultural sector as a whole in Lebanon. In the post-war years, given the relationship between Syria and Lebanon outlined in the Introduction, Syrian agricultural produce was sold in Lebanon at cheap rates that disadvantaged local farmers.

In spite of its rapid rise, the quarrying industry was not invulnerable, either. Even as it was emerging in Arsal, it faced an ambivalent future. Transforming private lands into quarries or, more widely, 'borrowing' communal land for this purpose, was only possible because state regulation was non-existent. After the war, the slow recovery of state institutions suggested that regulation was imminent. The decree mentioned at the beginning of this chapter was intended to reassign specific land for quarries while closing down those that were operating informally in the hope of restoring farming lands. This was still a matter for debate, but it threatened the many existing informal quarries and stone factories situated on Arsali land. It also flared up the risks of 'experimentation' and the vulnerabilities of rural livelihoods. Would the *tajruba* of quarrying fail? And would it fail *despite* people having resigned themselves to its harsh demands? At the heart of these questions lay mistrust of the state's intentions towards the Biqaʿ region in general and Arsal in particular. The following section elaborates further on how the uncertain nature of this law convinced Arsalis of their marginalization by the Lebanese state.

2   Livelihoods in the Shadow of an 'Evil State'

To understand why the state generated such mistrust, we need to place the quarry law within the larger context of regional inequalities in Lebanon and the particular milieu of unrest in the northern Biqaʿ. State neglect of rural areas was not specific to Arsal and went back a long way. Even before the 1975–1990 war, rural parts of Lebanon were undergoing what Salim Nasr considers 'a stage of decomposition and permanent crisis' (1978: 8) due to the dominance of the financial and commercial sectors.[6] After the war ended, capital was invested in urban areas, especially Beirut, at the expense of Lebanon's regions. In a bid to revive the economy, neoliberal economic policies were focused on enhancing those same traditionally strong sectors of the economy, and the

---

[6] Nasr argues that 'the relative share of agriculture in the Lebanese economy decreased from 20 percent of GDP in 1948 to 12 percent in 1964 to less than 9 percent in 1974' (Nasr 1978: 8).

attention paid to agriculture and industry remained negligible. In the hope of attracting direct foreign investment, Lebanese officials in the mid-1990s began to negotiate free-trade deals bilaterally, regionally, across the Mediterranean, and globally (with Syria, Arab countries, the EU, and the WTO respectively). These deals privileged foreign international relations and neglected domestic aspects of economic policy, including focusing on improving the public sector and advancing the agricultural and industrial sectors through incentives to modernize (Baroudi 2005). The repercussions of these agreements and policies culminated in protests by farmers across the country. The following is an extract from a statement made by the Agricultural Coordination Committee:

> freedom of trade between Lebanon and Syria or between Lebanon and Arab countries [was harming] Lebanon's interest, because of higher production costs … [that] result in unfair competition, leading to heavy losses for farmers and the bankruptcy of agricultural firms.[7]
>
> BAROUDI 2005: 207

By 2001 these protests were turning violent in some parts of the country including the Northern Biqaʿ, especially Hermil. Farmers burnt tyres, blocked roads, and emptied crates of local produce on the roadside. This mobilization highlighted the resentment felt by rural communities for the urban-centric planning that dominated successive post-war governments. An immediate response by the Lebanese National Security silenced the protests, however. Sami Baroudi (2005) suggests this swift intervention was due to the anti-Syrian aspects of the protests at a time when overt political expression against Syria was not yet tolerated. With the exception of these short-lived protests, the agricultural lobby was unable to mobilize the farming community into meaningful political action.

As I mentioned in the previous section, profits from fruit farming in Arsal were also disadvantaged by competition with Syrian produce. The growing Syrian control of the town's economy and security was held to be the direct result of Lebanese state negligence on the one hand and of its 'weakness,' as demonstrated through its acceptance of Syrian abuses in Lebanon, on the other. 'Where is the state?' (*wayn al dawlah?*), people demanded, echoing a question that is cited almost universally in the ethnography of Lebanon (Monroe 2016; Hermez 2015; Obeid 2015, 2010b). Lebanese state presence was episodic. Its absence was felt in its failure to provide services to which the residents felt they

---

7 This quote is taken from a statement issued on 18 July 2001 by the Agricultural Coordination Committee and published in Al-Nahar, 19 July, p. 6.

were entitled. When the state materialized,[8] it did so through its apparatuses of force, to quell a demonstration, to arrest citizens, or to protect a visiting high-profile official during parliamentary elections. Most resonant in the town was the belief that the Lebanese state turned a blind eye to Syrian transgressions, from the petty bribes that the *mukhābarāt* (secret service) agents collected from truckers to allow the flow of produce in and out of the town, to unlawful arrests and kidnappings in the highlands, and interventions in the local municipality council. By the time the dubious quarrying decree came out, mistrust was rife in Arsal, and this fostered scepticism about the state's interest and intentions in their town. These sentiments can be traced in the official statements made to the press in response to the law.

After a visit to Arsal and while the quarries were being closed, a sympathetic journalist reported the following,

> Whoever spends a day in the quarry will recognize perfectly well that this is not an occupation that pours *dahab* (gold). If anything, it pours *ta'ab* (weariness). The [labourer] seems like a gambler who may lose everything if he does not live up to the expectations of the land. It is an occupation of hardship, weariness, the sun, the wind, the cold and nakedness. The Arsalis have yielded to it as means for making a living, away from the path of smuggling and the hardship of migration. However, their government has not yielded and brought about its arbitrary laws that have nothing to do with environmental, social, natural, geological, or economic standards.
>
> AL-JAMR 2003 [my translation]

In his article, the journalist echoes the feelings of Arsalis that the state had dealt with Arsal in an arbitrary manner and, by doing so, deprived them of an occupation they had embraced despite the risks to their health, and sometimes to their lives. According to the newspaper article, the Ministry of the Environment assigned the most unsuitable land to be utilized for quarrying, in the northeast of Arsal. But the mayor clarified that

> [the northeast] is the farthest point from the [residential parts of the] town. There is no electricity, no water and no roads [there]. We are all

---

8 The state has been theorized as an 'ideological artifact', a 'cultural artifact', and as a general aggregate of ideas, practices, apparatuses, affects, discourses, and representations (Sharma and Gupta 2006; Abrams 1988; Mitchell 1991; Trouillot 2001; Steinmetz 1999; Aretxaga 2003; Li 2005; Nuijten 2003; Navaro-Yashin 2002).

aware that there has been historical conflict with our Syrian brothers over land ownership in that area where plots of land are intermingled. Why couldn't they [the state] have just adopted the areas where the quarries presently lie? At least the general prerequisites do exist: they are environmentally sound, there are no trees around them and houses are far away from them.

The mayor's statement can be interpreted in two ways. The first is that the government had 'arbitrarily' decided on a spot without a scientific survey of the land.[9] The second, which carries a greater suggestion of conspiracy, is that the state intentionally assigned a 'conflict zone' to make the lives of the Arsalis more difficult. The conflict refers to disputed plots of land on the border between Syrians, Arsalis, and two neighbouring Lebanese villages. One interlocutor explained to me that any plot left for a number of years in that area would be confiscated by the Syrians. He showed me a particular plot that was claimed by Syrian farmers and said that the owners tried to complain to the Lebanese authorities but their reply was that 'nothing can be done if the Syrians took it.'

The idea of regulating land-use is not entirely novel to the Arsalis. Prior to the 1975–1990 war, when transhumance predominated, the communal lands of Arsal were managed through a local system that broke down during the war (Chapter Four). Therefore, different land-users in Arsal were not against regulating land per se, as ideally this would have benefited everyone. Rather, they were worried about the manner in which the law was 'fabricated' (col. *mufabrak*).

Undoubtedly, the decree roused anxieties about the fate of illegal quarries on a national scale, and in different parts of Lebanon. In Arsal, the particular worry was that local quarry owners were not entangled in patron-client networks that connected them to the state, unlike the well-known names of big business men, and more particularly one notorious minister, who owned quarries around the country. By the late 1990s, the media and environmental organizations were fiercely attacking the Ministry of Interior for its corruption in issuing unlawful licences to people in power. The fear was that these big owners would manipulate state regulation to serve their own needs. With regards to the law under discussion, there had been talk of imposing a fee of 40,000 USD for a licence, which only wealthy owners could be presumed to afford. A state that was historically hostile towards Arsal, people reasoned, would either

---

9   Part of the problem of a 'scientific' survey of the land is that it will target residential areas of Arsal as well as land in the highlands. Over the years the town has been expanding on communal lands. It is unclear what the fate of offenders will be.

give their land away to other quarry owners – turning them into 'slaves on their own land' as was repeated to me – or shut down their operations without offering alternative livelihoods. I was told that representatives from Arsal had attempted to negotiate with the Ministry of the Environment[10] to change or reconsider central aspects of the *Mukhattat Tawjīhī* (National Master Plan) that were not to their advantage, as the mayor pointed out. A 'promise' was apparently made concerning the negotiations, but Arsalis were still waiting for the government to reconsider, albeit without faith.

The state's ability (or intention) to provide alternatives had only recently been put to the test in the surrounding Biqa' region, with disappointing results, when Lebanese and Syrian forces had attempted to eradicate the illicit cultivation of opium poppies and cannabis between 1991 and 1993. Residents of the Biqa' were highly critical of the discourse of 'alternatives' (*badīl*), be they alternative economies, farming, or development initiatives, that were imposed on the region but left families (and entire villages) without income, despite the considerable UN funding for the so-called Baalbek-Hermil Regional Development Programme. UN agencies acknowledged this failure in some of their literature; one report, for example, concluded that 'the eradication campaigns were not ... integrated into a wider programme to address the socio-economic conditions in the Bekaa valley and to limit the loss of income for the affected communities.'[11] While the reduction of the cultivation of these crops was cause for celebration in official and international circles, it was ridiculed locally as a failed exercise in development (*tanmiya*) and governance.

In the case of the quarries, people in Arsal worried that they, too, would be left without feasible alternatives, especially when no alternatives had even been proposed this time. State intervention was not an issue that people regarded favourably. A quarry labourer who was interviewed in the newspaper article quoted earlier, summarized in a couple of words what he felt the people of Arsal thought of their state. His wellbeing and that of his family, he was quoted as saying, would *only* come about when the state decided to 'turn its evil away from us.' In this case, state absence was less detrimental than its presence and intervention. This echoes what my interlocutors at the beginning of this chapter were expressing. The disgruntled Kamal told me that by depriving them of their livelihood, the state was symbolically executing the Arsalis. He invoked the proverb that best reflected their sentiments upon the decree – *qat'*

---

10   On matters relating to quarrying, mining and other industrial projects, the Ministry of the Environment shares a mandate with the Ministry of Interior.
11   See http://repository.un.org/bitstream/handle/11176/239028/E_CN.7_2002_6-EN.pdf?sequence=3&isAllowed=y.

al-aʿnāq wala qatʿ al-arzāq ('Cut the necks but never cut the livelihoods'). In other words, the deprivation of a livelihood is equivalent to, if not worse than, taking someone's life.

3 Contested Moral Economies

The imminent enactment of the law stirred up a need to defend quarrying, and this became a mobilizing issue in the local elections that took place later in 2004. But quarrying was not unopposed in Arsal, and the relationship quarry workers themselves had with this livelihood was complicated. Although several new occupations have emerged since the beginning of my field research in the 1990s, in what follows I address some of the moral questions that quarrying was posing as a new livelihood at this time, particularly in relation to the traditional livelihoods historically associated with an 'Arsali way of life.' The manner and speed in which this way of life was perceived to be changing was contested in different ways in the town and was central to how people evaluated their continuing 'experiments' and search for sustainable livelihoods. I am guided by Andrew Sayer's 'simple question' here: 'What are economies or economic activities for?' His answer is 'to enable people to live well. What else could it be for?' (2000: 80). This 'living well' is not contingent on economic gain alone, as advocates of 'embedded economies' have long argued (Polanyi 1957; Scott 1976; Thompson 1971; Booth 1993). In unpacking the complexities of living well, I follow Sayer's approach to moral economy, which he considers as a form of inquiry (like history, for example) rather than just an object of study. Like him, I would like to employ moral economy to look at the 'ways in which economic activities in the broad sense are influenced by moral-political norms and sentiments, and how, conversely, those norms are compromised by economic forces' (2000: 80).

'Living well' was a contested matter between advocates of traditional livelihoods and those who preferred some of the new ones, especially quarrying, which increasingly attracted a growing labour force. My aim is not to draw oppositions between tradition and modernity, nor to claim that the views discussed below constituted a rigid division between old and young. Rather, I wish to draw out tensions and points of contention, some of which were indeed articulated as generational differences in the context of unstable and uncertain political economies. One of the core assumptions relating to moral economy is that forms of production should contribute to 'proper and socially valuable outcomes' (Griffith 2009: 438). In the following, I discuss two main socially valuable outcomes that shaped conversations surrounding livelihoods:

the preservation of kinship and communal networks, and the relationship between human wellbeing and the environment.

3.1 Living Well through Traditional Livelihoods

Advocates of traditional livelihoods, no matter their age, valued the ethos of cooperation that defined agriculture in its two branches, *zirāʿa* (farming) and *ghanam* ([literally sheep] herding). They celebrated the seasonal exchanges that fostered reciprocity, conviviality, and sociality as described in the previous chapter. They also championed the centrality of kinship in the daily organization of these activities. By demanding labour and commitment for the sake of more than the self, agriculture was believed to enhance notions of sharing and mutual engagement. Yet, social norms and obligations are not stagnant. Michael Lambek argues that 'the story of modernity is, in part, a story of anxiety over the ostensible loss of kinship (as much as or irrespective of its actual loss)' (2013: 242). This anxiety was manifest in the way an older generation of herders laboured to safeguard pastoralism in the wake of its decline, not just as a source of income, but as a way of life.

One of the key values one herder, Abu Ahmad, ascribed to this livelihood was its ability to safeguard the family at a time when many changes seemed to want to fracture it. Pastoralism, he told me, '*bi-dubb* (col. contains) the family.' As part of a larger group, individual family members are continuously supervised. Because agropastoralism imposes demanding and time-consuming labour, Abu Ahmad reckoned that this made individuals too busy to engage in other dangerously perceived enterprises. The fact that shepherds live in the highlands most of the year, for example, prevents them from mingling with 'street boys' in the town, therefore maintaining their decency and avoiding trouble. Another herder, Abu Khalid, told me that during the 1975–1990 war the herd 'saved' his eldest son from the political parties that recruited young men and made them fight in the city. Fearful of their influence, the father sent his son away to the highlands where the herding unit camped. 'As soon as he arrived, he got busy with work, running after the sheep, feeding them, fetching water … he forgot all about these ideas.' He continued in a more facetious tone, 'If all the Lebanese owned sheep, we wouldn't have had a war!' Labour in this perspective is valuable not just for its economically productive worth but because it refocuses the attention of youth towards their families, giving them clear purpose, an offshoot of which is the avoidance of conflict.

The herds are believed to contain the women just as much as the men. It is perhaps too much of this 'containment' that gave herders a reputation for being more conservative than other families whose daughters may have had more freedom to study in nearby universities, work in paid employment, and travel on their own. 'The herders are strict; they do not allow their girls to go out unnecessarily into the street,' I was told by Abu Khalid's daughter, who had recently been scolded by one of her older brothers for flirting with a boy on the street. The street represents the space where girls might face the harassment of young men. Alternatively, it is the space where an unsolicited relationship could start when young men and women exchange amorous looks and smiles. If seen too frequently on the street, and without a clear purpose, girls and young women may be labelled 'loose' (*faltāna*). Because herding demands intensive collaborative labour and spatial distance from the town and its dangers for a substantial portion of the year, its advocates believe it to be a livelihood that strengthens family bonds and safeguards the moral integrity of its members by containing them and by making it possible – or rather impossible not – to spend long periods of time together.

The type of family labour and commitment that pastoralism requires, however, is becoming increasingly impracticable in Arsal. As life changes, young men and women feel constrained by the priorities that the herding unit enjoys over their own desires. My young interlocutors who belonged to herding families repeatedly told me that the returns from a pastoral way of life were hardly worth the *ta'ab* and *shaqā'* (tiredness and weariness) invested in it. This kind of sentiment became all the more salient when young men and women compared their tired lives with those of others in the town. As we will see in the following chapter, an individual's pursuit of marriage, education, and employment was often thwarted for the benefit of the larger herding unit. The supposedly socially valuable outcomes of family containment and cooperation were being questioned by younger residents who experienced these, in their extreme, as restricting and stifling their life plans.

In addition to the preservation of kinship and communalism, advocates of agriculture valued the relationship between farming and herding and the environment (*al-bī'ah*). The surge of quarrying had been causing alarm over its detrimental effects on fertile land. Fruit orchards close to quarries were suffering, and shepherds were faced with having to find new grazing routes for their herds. Land-use conflict became a pressing issue for the new locally elected council in 1998, and for its successors, as different users made claims on how natural resources ought to be utilized, bearing in mind future generations. This became particularly relevant in the context of a growing discourse about 'sustainable development' promoted by local NGOs and projects led by the American University of Beirut. The herders and full-time farmers I spoke with took pride in how their livelihoods respected nature and worked with rather than against it. As we saw earlier, fruit trees were considered to improve the conditions of Arsal's soil, not to mention bringing beauty to its arid landscapes.

In a similar way, advocates of herding felt that this livelihood respected the natural cycle, whether through the planting of cereal crops or by nurturing a harmonious balance between animals and humans.

As a natural space, the *jurd* (highlands) played a central part in how people understood and experienced the demands made by different forms of work. For example, when comparing herding with quarrying, Abu Ahmad assured me that while both occupations were demanding on the body, 'One shortens life, the other prolongs it; never in your life will you visit a hospital if you are a herder,' citing himself as an example to make his case. At that time, Abu Ahmad's health fluctuated on a seasonal basis. He had been diagnosed with diabetes, but the instructions of his doctor in Baalbak to eat less fat and sugar were scorned in a household where every meal was cooked with fresh ghee from his herds in the highlands. His daughters laughed at the idea of their stout father eating a bowl of cereal ('low-fat Corn Flakes', as suggested by the doctor) in place of his usual *tulmiyya* (a thick, homemade flatbread) or two filled with cheese, fried eggs, preserved strained yoghurt, and homemade cherry jam. Come summer, however, he would move to his highland house and spend weeks in the field, ploughing land, planting vegetables, harvesting wheat, and picking fruits, punctuating these activities with visits to his sons in their nearby herding camp. There he would check on the health of the sheep and sometimes take part in milking. He insisted that he felt more alive, at the age of 70, when being active, eating fresh food, and breathing fresh air. In the town, he felt reduced to a '*khityār*' (old man). Above all, he described feeling free in the highlands.

People often repeated the common saying *al-barriyya hurriyya* ('the wilderness is freedom') to me. In most contexts, they were commenting on the luxury of space in the highlands as opposed to the oppressive feeling of being crowded together in the rapidly expanding town. The *jurd* was lauded as a space of vastness in which people could indulge in 'acting naturally.' For example, people often commented on how, in the highlands, you could tear meat apart with your hands rather than worry about the use of cutlery, which belonged to the civilized space of the town and cities. This sense of emancipation from certain rules was sometimes stretched to allow other transgressions of behaviour. One May, a group of us received a *'awnah* invitation to shear sheep, as described in the previous chapter. When we arrived at the designated plot, Abu Ali, one of those who had invited us, an elderly, normally dapper, man, warned me: 'You will have to excuse whatever you hear from now on. The wilderness implies freedom and we have to joke to pass the day.' The ribald jokes that followed certainly pushed the boundaries of what was normally acceptable, and were met with giggles by the younger workers, who enjoyed seeing this man's mischievous side. All the participants agreed, at the end of an exhausting day, that what had made it tolerable, if not enjoyable, was the support and cooperation of others, the 'freedom' to laugh and joke, and the clean air and fresh food.

But if some romanticized the wilderness for its unbounded nature and for allowing a degree of malleable decorum, others criticized it for those very same reasons. The wilderness was conceived as being outside of modernity, away from civilization, untamed and uncontrolled. One shepherd told me 'Herding and the wilderness are no longer *dārja* (fashionable),' explaining that the *jurd* had ceased to be a desirable habitat to live in all year round. Even men who worked in the quarries returned to their houses in the town so as not to have to sleep in the wilderness. He himself was lobbying his parents to quit herding, to no avail. Many young men with off-farm jobs who might still enjoy an occasional leisurely trip to the highlands, usually for hunting or barbecues, resisted their parents' demands to help out with seasonal agricultural work. As Adnana explained to me, 'If you want to rest after a week of work, you want to sit and watch television, visit your friends and sip some *mate*, or take a trip to Baalbak. You don't want to go to the *jurd*! There is no rest (*rāha*) there.' She was explaining to me why her three older brothers who owned a garage together had refused to accompany us on a tree-pruning trip to the highlands on their day off.

'Living well', then, is a contested matter. Kinship-centric livelihoods that thrive on cooperation and reciprocity produce socially desirable outcomes. The question is to what extent and for whom. The same question can be raised in relation to the contention that being in the wilderness and engaging directly with nature produces better lives and improved states of wellbeing. These differing moral positions and disputes about livelihoods were at the heart of a transforming economy. But how were these questions unfolding in the new quarrying sector?

3.2   *Quarrying: Prosperity or Sentence to Hard Labour?*

Like other livelihoods, quarrying attracts conflicting moral opinions. At one level, quarrying represented the ability of Arsalis to create livelihoods out of a situation of lack. In their accounts of experimentation, my interlocutors boasted of their resilience and resourcefulness as they narrated how they strove – remember 'God's servant must (*yas'a*) strive' – to create new opportunities out of the rock. The socially desirable outcomes of this development were felt at many levels. On a national level, the growing industry made Arsal a 'centre' for the distribution of stone to different parts of the country. By the end of the 1990s, the town had made a name for the particular quality of its stone and its expertise in stonemasonry. Some Arsali men were finding work across the country. For example, Naji was hired as part of a team of Arsali expert stonemasons to work for a year in one of Lebanon's prestigious ski resorts in the mountain village of Farayya. He told me how much he had learnt during that period and appreciated the exposure he had to new styles of architecture while working on the luxury villas. Equally, he felt proud that Arsali stonemasons were shaping a trend outside the town by exporting the local aesthetic of rough untreated stone exteriors and *kuhl*, the black lines painted between individual stones. Later, Naji took on new projects in different villages in South Lebanon, where new markets were developing in the post-war years as members of the Lebanese diaspora in Africa invested heavily in property back home. Although his work took him away from Arsal, Naji's trade enabled him to support his growing family still based in the town. On one of my recent visits to Naji's house, I found he had built a chimney in the main room, copying one he had seen in one of the resorts.

On a regional level, the industry is thought to have enhanced the town's economic standing within the Northern Biqa', particularly in comparison with some of the poorer neighbouring villages in the valley that rely solely on farming. But it may have also created some resentments. A truck driver who transported rocks from the quarries told me he believed that that the wealth quarrying is perceived to have generated in Arsal has created a feeling of 'envy' (*hasad*) in the Northern Biqa'. This was complicated by the political developments that polarized the country after the assassination of Prime Minister Hariri in 2005 and the Israel war that followed in 2006, when Israeli warplanes shelled the border highlands. People described an eerie emptiness in the streets throughout the month of July, but the quarries continued to work, in spite of the danger. Smugglers in Arsal also carried on their activities, transporting diesel to hospitals in the region. The same truck driver complained to me that these 'heroic' actions were misinterpreted in view of the region's mounting sectarian tensions. Locals in the neighbouring Shi'i town of Labweh erected a street sign leading to Arsal that read 'This way to Israel' (see Chapter Eight). The driver reasoned that this slur on Arsal's residents and their own history of fighting against the Israeli army during the 1982 incursion occurred because Arsal was 'standing out politically and economically.' He explained that by expressing anti-Syrian and anti-Hizbullah sentiments, the inhabitants of Arsal had isolated themselves in the Biqa' region. Yet through their resilience and creativity they had been able to defy circumstances and carry on with their livelihoods.

Unlike agropastoralism, which ties labourers to the web of the family and the herding unit, and farming, which is seasonal and unsustainable on its own, quarrying promises a measure of income stability. The industry itself is stratified depending on where one sits within its hierarchy: a quarry owner reaps different rewards than a freelance truck driver or a day-labourer. Although the perceived affluence the driver referred to does not apply to the majority of Arsal's residents, a few large, newly built multiple-story villas surrounded by ostentatious four-wheel drive cars, give ample evidence of material prosperity. They usually belong to quarry owners and a few stonemasons. It is not uncommon to see unequal wealth within the same Arsali family. For example, Basil was born and bred into a herding family. But he built a career in stonemasonry and was doing well by the time he was in his mid-twenties. He was able to build a huge five-bedroom villa on the fringe of the town, displaying his own beautifully carved stonework, and moved into it with his new young bride. By contrast, his brother who shared herding responsibilities with his parents could never afford to build a house on his own. After much negotiation, his parents built a floor with two rooms on top of their house so that he could marry his cousin, also a herder, to whom he had been engaged for more than six years.

Day-labourers were not as privileged. The regularity of income, which amounted to 10 USD per day in 2003, made quarrying jobs attractive, but the risks were high. Almost all the quarries operated informally without an official licence; hence quarries did not observe the labour laws, which meant labourers were medically uninsured. It was not unusual to hear that a young man had died because explosives had been mishandled, and limbs were frequently lost. News came to me that Yahya had had to have his foot amputated following an accident. It was the first time I knew the person involved, at least at second hand through my friend Fahd, Yahya's brother. I arranged to meet Fahd in Beirut, where Yahya had been hospitalized because medical care there was thought to be more advanced than in the Biqa', though much more expensive, to find out how the family was coping. He told me that Yahya's three brothers and their wives, as well as his father and mother, had accompanied him to the city. Fahd described how devastated his mother was: 'You could hear her wailing from the end of the hospital ward.' She feared her son's life was wasted. Determined to show strength of mind, Fahd told me that he had reprimanded his mother for her 'over-reaction.' Turning to Yahya, he reminded him that he was a *zalama* (col. man). 'It's nothing serious,' Fahd insisted. 'The foot was amputated,' he told me, 'so what? At least he can still work, it could have been much worse.' 'People like us,' he went on, 'what [else] can we do?' It was unclear whether this question was to do with the fatalistic approach to making a living, given the belief that everything is *min Allah* (from God), or whether it hinted at Fahd's fear of what will become of his brother in a social climate that is not kind to the disabled.[12] But Fahd was drawing on an idea of masculinity that expects men like his brother to face the risks and endure the consequences. The idea that quarrying epitomizes masculinity in the shape of physical strength and endurance is reinforced by the fact that women have no place whatsoever in the industry.

In an environment where jobs are scarce and livelihoods uncertain, these risks do not deter young men from joining the labour force. During my research period, several boys who were not doing well at school dropped out to find jobs in the quarries. This broke the heart of many a parent who aspired for a better future for their children, one attained through literacy and education rather than hard labour. Zuhayr was one such parent who had high hopes for his sons and daughters. He promised to send them to Beirut, or even abroad, to pursue university degrees if they succeeded in completing their secondary school education. But his eldest son Mustafa had already failed his secondary school exams twice in a row, and, to his father's dismay, became a labourer in a quarry. As an environmentalist activist, Zuhayr was fundamentally opposed to the spread of quarries around Arsal, and felt that his son had just signed away his health. He was frustrated by the lack of opportunities for the young.

> You ask me why I have been feeling depressed. Look at Arsal, it has become a large prison with a life sentence for *ashghāl shāqqa* (hard labour). We can no longer work in Beirut. Even in the 'misery belt' (the suburbs) you have to pay 200 and 300 USD for rent. The Arsalis are trapped.

Mustafa had just come home after being caught by his employer angrily kicking the wall and swearing at God for bestowing such a miserable life on him. An experienced man who understood the challenges of this work, his employer ignored the young man's outburst and allowed him to let off steam. Zuhayr exclaimed that quarrying does make one want to fight with his God (*yukhāniq rabbih*). Both son and father were rebelling against the Arsali mantra that you must accept your fate because 'everything is from God.' Mustafa was passionate about the civil defence, the Lebanese public emergency response service, and had been volunteering twice a week as an ambulance driver. He had been undergoing first aid training in the hope that one day he would be employed on a full-time basis. But the prospect of a vacancy was dim. In the meantime,

---

12   One of the NGOs in the town worked on disability issues. From my conversations with its members and others in Lebanon, it appeared that disabilities were met with particular shame in rural areas.

he needed to work to help his father. Without much choice available, he had turned to the quarries. But Mustafa found 'God's will' too demanding and ended up quitting and then going back to his work at the quarry several times.

One evening, Mustafa's friend was visiting the house. While we had tea, Mustafa's mother bantered with her son who apparently had marriage plans. 'Mustafa is in love (*bi- hubb*),' she teased. 'What about you?' I asked his friend, who replied with a question. 'How old do you think I am?' he challenged me. '19?' I replied, assuming he must be about Mustafa's age. He was surprised that I had guessed correctly and complained that quarrying had added years to his life: 'Look at my face: it has wrinkles now. My body aches and I feel 30 years older than my age. Only two years ago I used to stay up late and enjoy spending time with my friends. This job [quarrying] has destroyed me.' For day-labourers on the lowest rung of quarry work, questions of health, wellbeing and exploitation ran parallel to the need to keep this industry alive as a potentially viable experiment – the more so since it was happening in a transforming social environment that was placing increasing pressure on young men to bear the brunt of production as people moved away from family-based livelihoods. These transformations in gender and kinship will be explored in more depth in the next chapter as I look at the challenges facing pastoralism in post-war Arsal.

CHAPTER 4

Pastoralists: Living the Past in the Present

One afternoon in the herders' camp, Nadia and I were engaged in the arduous task of churning butter. Shaking the *darf* (the goat skin into which yoghurt is poured and left overnight) was usually a daily task that took her, working on her own, over an hour until the butter formed. But work had been so intensive the week before, neither she nor her brother's wife, Suad, had had time to do it, so Nadia and I now found ourselves having to shake three *darfs* in a row. At first the process seemed easy; we each placed one palm on the middle of the goatskin and gripped the outside edge tightly with the other hand to shake it. The rhythmic slurping sound of the liquid was soothing. But by the time we started on the second *darf*, our arms were starting to ache and our knees had become sore from kneeling. Nadia began to sing, a guaranteed technique, so she said, for making time pass. At the top of her voice, she belted out a song by the famous Lebanese singer Najwa Karam:

> Don't cry, oh roses of the house
> Keep singing in his absence
> But when he returns from his journey
> Keep decorating the doors of his house

As Nadia sang, her grip on the *darf* loosened and the tune was abruptly interrupted by the *darf* exploding as it hit the ground, leaving Nadia with splotches of yoghurt all over her face and shirt. Startled, I jumped backwards, only to plunge my elbow in a bucket of hot milk hidden behind me. Much laughter ensued. 'It's good for your complexion,' I teased her. 'You must take care of your beautiful looks now that you are going to be married soon.' 'How can I be sure of that?' she replied grimly. 'My mother does not want to let me go.' Suddenly Nadia looked sad. The last *darf* went by quickly as she told me her story and we forgot about our sore arms and knees.

Nadia fell in love with Mahmud, a freelance stonemason, when she was 13. Seven years later, despite her mother's disapproval, they became formally engaged after he asked his uncle to intercede with her father. Her mother's objections that she wished for a 'better' suitor for her daughter were not entirely without self-interest. Nadia belonged to a pastoralist household. Her father and his brother jointly owned some 1000 head of sheep and goats. Three of her brothers tended the herds and lived permanently outside the town. Nadia was a central member of the household and her departure from it would have repercussions. She and her younger sister Laila had dropped out of school when their older sisters married and took care of all the household tasks together, thus sparing their mother, who had always taken on a significant share of the herding work.

From a labour perspective, Nadia's importance was that she was experienced and skilled not only in looking after their town house but also in tasks related to herding. Every year between April and November her parents sent her to the herders' camp to help her brother and his wife who, after ten years of marriage, remained childless. Her other brothers, the shepherds, were not yet married. Until one of them found a bride who could replace Nadia's labour, her departure would leave either Suad or her mother without help. If, on Nadia's marriage, her younger sister were sent in her stead to help her brother's wife, her mother, who claimed she could not manage at her age, would end up alone. If Suad were left without help, she would be unable to cope with the combined tasks of milking, making dairy products, and tending the tent-house and the three men. Nadia's lot, then, was tied to the interests of a web of other household members. When she tried to justify her own agenda, the moral rhetoric of the family's wellbeing – and hence her obligations towards it – seemed to triumph.

While she waited for a way out of her predicament, Nadia imagined a future in which she could follow her own dreams and aspirations. She was interested in flower arrangement and hoped that after she married, she might get some training and maybe even open her own flower shop. In those dreams, she wanted to focus on herself, her husband and two children, for Nadia insisted to me that she was going to break with the Arsali predilection for large families – she herself had eight brothers and three sisters. 'My life will be as far away as possible from *ghanam*' (literally sheep; here she meant the life of agropastoralists). Echoing the views of her sisters and some of her cousins, she considered her family's lifestyle to be increasingly impracticable, imposing labour demands and commitments that seemed to be 'remnants of the past,' unrealistic and unfitted for modern times. The tensions between Nadia's desire to pursue her own path and her family's efforts to 'contain' the family by seeking to ensure the commitment of individual family members to their joint livelihood (see Chapter Three) reflected a key challenge to herding in post-war Arsal. One of the most profound experiences of social transformation in the town has been the decline of agropastoralism. My interlocutors spoke about their present as deriving from an agropastoral heritage engrained in a distant past, even when about 10 per cent of the population was still engaged in full-time herding, and had one of the highest livestock counts in Lebanon (Hamadeh 2002). The life of herders was often romanticized, evoking an innocent past that celebrated togetherness and cooperation and admired the ways in which herders had resisted the changes that had swept across the rest of the town and region. More often, though, I heard scepticism being expressed over the harsh social and ecological conditions in which pastoralism operated.

A wealth of studies on pastoralists has questioned the assumption that mobile pastoralism is necessarily dying in the Middle East and North Africa and has shown the various ways in which pastoralists have adapted to changing environments (for example, Chatty 2006; Mundy and Musallam 2000). I have no wish to treat the practices and strategies I discuss in this chapter as if they were set in an unrelatable past, in spite of the local discourses that make this claim. Rather, my aim is to capture the tensions emerging out of a context of social transformation in which ideas about being modern in rural Lebanon challenge the tradition of agropastoralism and the life it imposes. I therefore focus on the different levels of pressure confronting what remains of the livestock rearing system in Arsal. The dilemma of Nadia's family is not peculiar to the herders of northeast Lebanon. There is an argument that labour is a major limitation in all pastoral systems (Bonte and Galaty 1991), but this is not the only challenge that pastoralists face. Herding strategies often depend on a combination of the household formations that make labour available and the environmental and ecological factors that shape resource conditions (Coppolillo 2000; Butt 2011). Pastoralists in different settings deploy varied strategies to deal with pressures that are often caused by social, political, and ecological conditions. Indeed, these are the same factors that influence the strategies employed by Arsali pastoralists who have had to cope with a variety of transformations affecting their sustainability: political changes between the Lebanese and Syrian states that have made the traditional seasonal movement across the border difficult, if not impossible; the changing landscape of the Arsali highlands where pastureland is diminishing as quarries and farmland expand on to communal lands; and changing family forms that are rendering the pastoralist household undesirable for the new generation. At the heart of all these factors lie the transformations in gender relations and the emergence of the model of the modern *murtāha* (comfortable) woman who is often contrasted with her exhausted (*shaqyāna*) herding ancestors.

1 Transhumance and Political Change

Pastoralists in Arsal have always depended on moving seasonally with their livestock to find pasture (transhumance). Until the 1970s herders followed the same seasonal routine. Between April and November, pastoralist units camped in Arsal's highlands. The flocks, led by a number of shepherds, moved gradually from the lower lands, which were grazed first, up to the fields in the higher lands. Between December and May, herding units crossed the border to an area in Syria known as the Hamad, near the northern Lebanese–Syrian border, where they spent the winter with their flocks. While the Syrian authorities regulated the Hamad, the communal lands of Arsal were managed through a local system known as the *hima*,[1] which partially protected the middle and the highest lands from the beginning of April to the end of May each year in order to preserve rangelands. Specific plots of land were assigned to individual flocks for pasturage. Other herds were prohibited from grazing these lands. The same process applied to water resources: shepherds used only four springs and accessed water only every other day to conserve water (Baalbaki 1997). Watchmen were assigned to oversee this as part of their communal duties. Pastoralist strategies, including organization and seasonal movements, are contingent on the political context in as much as they are determined by climate and ecology (Marx 2006). Political flux within the town and at a national level increasingly impinged on Arsali herders' strategies. The *hima* system was disrupted in 1964 just after the local elections and the ensuing violence. This was complicated by the outbreak of war in 1975 that made land-use management impossible. Until the late 1980s herders were still crossing the border, moving to the Lebanese highlands in summer after spending the winter in Syria. But by the early 1990s, after the end of the war, border control became stricter from the Syrian side. Bureaucratic measures made it more difficult to cross the border and the seasonal movement to Syria proved gradually impractical and less lucrative, as a herder recollected:

> The Syrians used to prepare a *bayān* (document; permit) that would allow herders to cross the borders. We were required to take the flocks to a village called Dayr ʿAtiyya where the authorities would mark each animal. It was very clear that we had to report all our animals to customs because the officer counted them. On our way back from the east, after the sheep had delivered their lambs, we had to pass by the customs of Hisya. There, an officer would check the information in the *bayān* and count the males, females, ewes, rams, sheep, goats; whatever we had. All was written on paper. The minute we wanted to come back, we had to have a middleman

---

1 See Bocco (2000) for a discussion of the *hima*, a traditional system of range management in the Arabian Peninsula. It is a system that aims at controlling grazing in selected arid or semi-arid land, where nomadic grazing is the predominant system.

> who was to mediate between the buyer and us. There is a lot of [meticulous] inspection for the *bayān*. We are not allowed to sell our sheep; we cannot take bids for the sheep. It is a never-ending story, a headache. The Syrian authorities are becoming tighter with us.

My interlocutor's description suggests that herders were uncomfortable both with the hassle of the new bureaucratic procedures and the feeling of being monitored. The problem for them was not the sites they had been using for pasture. In fact, the herding families I worked with spoke nostalgically of the herding conditions in Syria. Because of the abundance of pasture, labour arrangements were simple and families spent less on fodder for the winter. As my interlocutor emphasized:

> We were very comfortable in the Hamad. There was a water tank that came every day and gave us water for a fee ... But we stopped going there because Syria is very difficult. Otherwise, the Hamad land is beautiful. In good years the fodder would last us for two years. Now the fodder we buy is hardly adequate for the sheep. We have already bought three truckloads of straw, ten tons of wheat, 20 tons of barley, four tons of dry bread, and 40 tons of bran and we are afraid they might not suffice for winter alone. And we still owe the salesman around three million [Lebanese pounds, equivalent to 2000 USD].

The changing policies at the border made Syria less accessible in the post-war years. As a result, herders were compelled to find new pasture and resort to other migratory patterns. By then, the Arsali highlands had become much more restrictive. On top of legally owned land, many portions of communal land had been unofficially expropriated, as people converted them to fruit orchards and quarries. This increased land fragmentation and substantially reduced pasture and grazing areas, leading to a noticeable decrease in the numbers of flocks (Deek 2003). By the early 2000s, most herding families were moving seasonally around the Biqa', hardly using the land of Arsal. Some opted to graze Arsal's highlands from June through November but searched for alternatives outside of Arsal in winter. One destination was Mount Lebanon, which, in spite of its cold temperatures, was suitable for over-wintering flocks. The challenge in these arrangements was the affordability of land rent at a time when fodder made up the majority of herders' expenditure. This, indeed, is one of the main reasons why moving to find seasonal pasturage is used as an adaptive strategy by different pastoralists (Abu Rabia 1994; Scoones 1994; Niamir-Fuller and Turner 1999). Given the shrinkage of pasturelands in Arsal, the costs of moving seasonally in terms of rents and the price of fodder were becoming more difficult to bear.

Nadia's herding unit was struggling with just these same costs. The shortage of pastureland and the difficulty of finding 'good' land were affecting the type and amount of labour required of household members. But to understand these constraints, it is necessary to describe in more detail the herding unit to which Nadia belonged.

2   Spatial and Human Organisation

Arsali pastoralists rely on cooperative herding: they combine the flocks from two or more households and share the labour of pastoral production. This arrangement is known as *khalt* (literally, mixing). As my elderly interlocutors explained, until about the end of the 1980s *khalt* partnerships were widely used, but took a slightly different form. *Khalt* did not always take place amongst kin, although kin tended to be favoured. In all cases, the partnership was subject to a customary understanding, known to Arsalis as *Sharīʿa Muhammadiyya* ('the Muhammadan Law' [Muslim Law]), which was binding for seven years. If a partner was unable to contribute labour, because his household was too small or for any other reason, he was expected to provide twice the number of animals. The partner with the set-up to do the herding contributed half the number. At the end of seven years, if no change had been made to the partnership arrangement, the herd was divided equally between them. By contrast, the herding unit that I observed between 1998 and 2005 combined a cluster of households that shared the herd and labour equally, with defined tasks for each, and without a fixed time frame. Each family marked their own sheep, and while some activities, such as shepherding and sheep shearing, were undertaken jointly, milking of the family herd (usually done by the women) and most other household tasks were carried out separately. Given that the unit's households were very close in terms of kinship, they often helped each other out in all cases. The viability of these arrangements depended on the full participation and cooperation of all the members, including children (Chatty 1980; Abu Rabia 1994). They were driven by an ethos of *taʿāwun* (cooperation) and a collective commitment that prioritized the herd and the herding unit.

Until the mid-1990s, Nadia's father mixed his herds with six of his brothers, and they all camped together in the highlands of a neighbouring village. As land became scarcer, the brothers found it easier to manage if they split into smaller units. Three brothers kept their herds separately and the other four made two pairs; one of these was Nadia's father and a brother. When I spent a summer in the highlands with Nadia's family, the herding unit consisted of two main households, and they camped next to one of the families that was herding independently. Nadia's brother and his wife lived in one tent-house, which was about 500 metres from her uncle's. Inside, the tent-houses resembled the town houses described in Chapter One. The main living area was carpeted and had 'Arabic' divans for seating, with a *yok* that stored foldable mattresses. A curtain separated this part of the tent from a smaller area that was used as a kitchen, with cabinets that stored cutlery, crockery, and cooking utensils. Families normally cooked on a small portable gas stove. The most noticeable difference between the tent-houses and town houses was the absence of fridges and washing machines. Once a week, the women hand-washed a pile of clothes. A small tent was erected at the back of the main tent-house. It was usually used for baking bread, boiling milk, and processing dairy products.

Herding households were spatially divided according to the season and the needs of the herd. Generally speaking, while some members went back and forth between the town and the highlands, others had to live outside the town all year round to accompany the large herds. In Nadia's household, there were four permanent herders (her three brothers and sister-in-law) paralleling another four in her uncle's house (father, daughter, son and his wife). These two households moved together as a group twice a year. In winter (December to April), they rented a plot of land in the lowlands, which had two small concrete houses, one for each family, and a large concrete enclosure for the flocks. At this season of the year, four shepherds were required to do the work. The flock was divided into two sub-herds, each managed by two shepherds. The three women looked after the domestic work and took care of milking and feeding the household animals. Meanwhile, other members of both families remained behind in the town, Nadia among them.

Towards the end of April, the permanent herders moved to the higher communal lands where they erected two tent-houses close to each other and stayed till mid-November. With the arrival of spring, when the herds had to be milked three times a day as opposed to once at other times of the year, Nadia and another unmarried woman joined the herders to help out with the extra load. When school finished in July, the children in both Nadia's and her uncle's households, along with her uncle's wife, joined the summer camp. The herds were divided further into four sub-flocks, each with two shepherds who would take a different path every day to the pasture and slept away from the tent-houses in the open air.

In summer, Nadia's father, mother and two brothers stayed back in the town to take care of jobs there. Her parents dealt with the agricultural harvest

FIGURE 9   A herders' tent-house

between July and September, usually asking for help from the herders in their camp (Nadia's parents owned cherry and apricot orchards, vineyards, and fields of chickpeas, wheat, barley and lentils). As for her uncle's house, nobody stayed in the town after the schools closed.

When school resumed in October, those who had joined the summer camp in June went back to the town, leaving the permanent herders behind. By that time, the herders had usually secured a winter camp and they moved back to their winter plot in the lowlands again at the beginning of November. Sometimes, all the family members returned to the town in November to rest while the shepherds stayed with the herds on the outskirts of Arsal. Although these households moved two or three times a year, they did not necessarily go to the same plots of land each time. As mentioned above, finding suitable land was becoming complicated; the choice of land affected the wellbeing of the flocks and their keepers. Moreover, the unavailability of land had direct repercussions on the spatial division of the herding unit, as well as the division of labour. The predicament that Nadia expressed at the beginning of the chapter was directly tied to these problems. The following section unpicks these complexities and shows how the availability of land is intertwined with labour problems and the wellbeing of both herders and their animals.

3 Herding Dilemmas

The permanent herders were in the summer camp as winter approached. November had already started and it was time to move. Deceived by a week of pleasant weather, they decided to take advantage of the remaining green pasture to try and save on fodder, thus postponing moving to the lowlands. But one night a storm flooded their tents and forced them to pack up everything and flee to the town. Suad, Nadia's sister-in-law, told me that they were up all night putting everything in boxes and loading them in to the truck, all in windy and wet weather. The prospect of spending a couple of weeks in the town was appealing since Ramadan had started and being with the rest of the family seemed like a good idea. As the women stayed in their town houses, the four shepherds remained in the lowlands with their flocks. The first lambing was approaching, however, and the family were under pressure to sort out the winter camp before December. By then, the few available areas of land had already been taken by other herders. Their choices narrowed down to two: a plot of land they had been renting previously, though they were not particularly satisfied with it because the sheep had contracted a disease there, or a better spot on a holding that had a large, newly-built enclosure. Evidently, the latter was better for the herd – but it had only had one small house, which could accommodate one household but not the full herding unit. On the eve of lambing, the wellbeing of the sheep was the priority, which meant that one household had to rent the older, less desirable plot. Since Nadia's uncle's was the unit with the most lambing experience, his household took the good spot where the female herd would be kept for lambing, while Nadia's brothers had to move the male herd to the less desirable plot.

Although the decision seemed the most practical one under the circumstances, it gave rise to a number of dilemmas that caused much anxiety before the move. First of these was the fact that, since the households were now being separated, Suad would be left alone all day long in the middle of a vast empty stretch of land. The shepherds would be busy supervising grazing throughout the day and, with this arrangement, her husband would have to spend the day on his tractor travelling between the two plots, transporting water and feed. Suad worried about how she would endure that winter. Her concern was not the amount of work, as she was used to taking care of everything on her own. Rather, it was the social aspect of having to spend so much time alone, and the safety factor. To allay her fears, one of the shepherds had to remain near the house in order not to leave her alone. This was not an ideal arrangement, not least for the herd because the animals had to graze in a limited spot.

Nevertheless, a shepherd was nearby if Suad needed him for any reason. This made her feel slightly more secure, but still ill at ease:

> Have you ever heard of a woman staying alone like that in the middle of nowhere? The boys are always out with the sheep and my husband is on his tractor. What if something happens to me? I am not comfortable and we are not used to this. Usually, we spend winter together [when the workload is reduced, with her mother and sisters from the other household], take breaks during the day, sip tea and even spend the evenings together. I feel very lonely here.

The other household was also disadvantaged because only two shepherds were available for lambing. Whereas in previous years all four of them would spend nights with the sheep, that year only two of them were on hand to do the job. Lambs born during the day were brought to the camp by the shepherds and left there, along with their mothers. The changes that year did not totally jeopardize the organization of the unit, but they made it more difficult for the shepherds as well as for Suad and her husband, who spent his days on the road between the two camps. This is just one example of how settling for available land in a context of shortages can negatively impact the herd and the herders, their spatial and labour divisions, and their sociality.

Unsuitable land may also affect the amount of labour needed. When Nadia's family camped in neighbouring villages or even in the Syrian Hamad, where there were few orchards or quarries to negotiate, one shepherd could handle all 500 animals on his own. He did not have to run around from place to place to keep the herd from wandering and had more time for leisure breaks. But when land is restricted, the task of shepherding becomes more complicated, requiring more energy and effort. Camped in the higher *jurd,* Nadia's herding unit was finding grazing increasingly difficult due to the gradual encroachment of the orchards. The shepherds could not afford to make any mistakes; they had to be vigilant about monitoring the herd, especially the 50 uncontrollable goats. Farmers were becoming impatient with the animals devouring their saplings and trees, and some even started taking extreme measures such as laying poison on the boundaries of their orchards. Others went to the shepherds' fathers to complain about the herds destroying their trees and asked for compensation. These incidents were the cause of embarrassment both to the fathers and to the young shepherds themselves, not to speak of the economic loss. Forced to work harder and harder, Nadia's brother grew so thin that his parents became alarmed and hired a Syrian shepherd to help him out until his younger brother had completed his academic year. Although the Syrian shepherd was experienced, it took him one day to decide that 'serving in prison was easier than herding in Arsal's land,' and left. Nadia's brother was not surprised because, as he put it, 'If it were not my herd I would have run away as well! This boy thinks our land is like their Hamad, a valley where the shepherd lets go of the sheep and sips tea all day long. No! It is different here, all these trees, you have to prevent the herd from entering orchards.'

Nadia's unit had found themselves in a vicious circle that year. They could not afford to rent land in summer (which would have cost them about 13,000 USD) as well as in winter. So they had settled once more for the highlands of Arsal, where shepherding, made ever more difficult by the spread of orchards, required more than the available amount of labour. In the midst of a tense discussion about how to help the young shepherd, Nadia's father managed to joke with his wife: 'I was right to want more children. I told you our 12 children would not be sufficient!' The suggestion that theirs was a small family caused laughter and lightened the atmosphere. But his joke brought home the point about shortage of labour. In the context of increasing pressure to find land, and the consequent tensions between safeguarding the wellbeing of the herd and individual household members, the herds came first, often at the expense of their keepers.

4 Conflicts of Interest

The relationship of household members to their labour evokes Claude Meillasoux's classic question: 'Who works with and for whom, where does the labourer's product go and who controls the product?' (1972: 98). The Arsali herders had their own answer to these important anthropological questions: 'We herders work for the sheep, so we follow them and their needs. The sheep follow the *miryāʿ* [leading ram].[2] The ram follows the donkey and there you go. All this hard work, it turns out we are all working for the donkey!' While told in jest, this sarcastic explanation reveals the internal dynamics of herding households and the tensions brewing behind individuals' overt commitment to the 'containment' of the family. The sanctity of the family represents a cultural ideal that has various articulations, not just among different communities in Arsal and Lebanon but also across the Middle East and beyond. Yet,

---

2 In Lebanon, Syria, Jordan and the Negev, herders choose a ram to be the leader of the flock. They attach a bell to its neck and never shear it, so that it stands out from the rest of the flock. When it is time to graze, the 'leader' follows the shepherd's donkey. The flock follow in lines behind the sound of the bell.

FIGURE 10   Young shepherd bottle-feeding a lamb

how individuals relate to their families is concealed by the power and purchase of this ideal. A unitary view of the family tends to eclipse conflicts, divisions, and tensions, thus flattening social relationships and obscuring the workings of power within households and families (Guyer & Peters 1987; Harris 1981; Whitehead 1981; Creed 2000). I draw on Anika Rabo's suggestion that we investigate how people '*do* family' (2008: 131),[3] a distinction she makes from '*talking* family,' which reflects 'how ideals about the family are verbalized' (2008: 131). The focus on 'doing' helps us understand how individuals do family *together*, even when this exercise may not always be characterized by cooperation. In the following, I explore the tensions between the local discourse of the family as a 'haven for unity and solidarity' and the lived reality of the family as an arena for obligations that generate inequalities and conflicts of interest.

Thus far, we have seen more than one case – Nadia, her brother the shepherd, and Suad – where individual interests were subordinated to the demands of the larger collective entity of the herding unit – in fact, merging with the collective epitomizes the very rationale of pastoralist life. Nadia's aunt elaborated on this through one of her proverbs: *hutt ra'sak bayn al ru'ūs wa qul ya qattaāʿ al ru'ūs* (Put your head among the heads and call on the beheader). This requires the individual to join the group, in this case the family or the household, regardless of the severity of the consequences, even death itself, so that hardships are faced collectively. Indeed, such a saying may exaggerate the extent to which a household is unified. Sylvia Yanagisako warns us that 'the assumption that those who live together share a collective motive … obscures the social process through which individuals negotiate the relations that give form to households despite different and sometimes conflicting goals, strategies, and notions about what it means to "live together"' (1984: 331).

In some pastoralist societies, the sheep may 'symbolize the solidarity of the whole household' (Abu-Rabia 1994: 2), mainly when every member has a vested interest in the herd. The Arsali case is a bit different. Even when all household members were working for the herd, they were not homogenously after the same end. Households in Arsal organized around the wellbeing of the flock as we saw earlier, but the herd unified members of the house from an organizational perspective. Rather than being a symbol of solidarity, for my interlocutors the herd came to represent a binding, restrictive component for different members of a household who had conflicting interests. In order to demonstrate this, one might start by answering Meillasoux's key questions by looking at access and control of resources in herding households.

---

3   Rabo draws on Margaret Nelson's 'doing gender,' 'the interactional work and activities through which connection is created and rehearsed' (2006: 782).

In Nadia's household a co-requisite for ownership was *shughl wa ta'ab* (labouring and tiring) over the flocks. But only male *ta'ab* seemed to count since five of the eight brothers, together with their father, had shares in the herds. The fact that they had *access* to the flocks by no means implied that they had *control* over them. Although different family members of all genders and ages usually took part in negotiations and discussions about most pastoralist-related issues, there was an acknowledged hierarchy of age, gender, and experience. The two younger shepherds reported to the two older. They made decisions relating to daily grazing routes. Nadia's eldest brother made decisions about buying fodder (quantities, timing, and composition) and feeding locations, to which he also transported water for the herds every day. Buying and selling animals was a much more serious issue that had to be approved by Nadia's father and uncle, the eldest of their households. Similarly, where to camp each year was a matter for approval by the eldest in the unit. Though all the males had a right to their property, they could not claim it without valid reason. How and when this property was transmitted determined the survival and reproduction of the pastoralist system. Even individuals labouring with the herds, therefore, could be deprived of their perceived rightful due in return for their labour.

For over two years Nadia's brother Hasan, one of the shepherds, had wanted to propose to his cousin, a permanent herder in the same unit. The two of them had grown up together and spent a substantial amount of time with each other, leading to the development of shared romantic feelings. His parents approved of the idea of his marriage for it could not have been more suitable: it would have kept them both within the unit, it was a preferential cousin marriage, and the two desired each other. Yet in order to marry, Hasan, like any good Arsali suitor, needed to have a house, or the capital to build one. His parents kept postponing his marriage project to benefit 'more pressing' issues, such as securing fodder for that and future years. That he was 'still too young' was really an excuse to put other needs first. He, on the other hand, was becoming impatient and felt that it was his right to access his share of the family assets. More than once, he angrily threatened to split off his share of the herd and sell it, which would have given him the money he needed for the marriage. Although a son could claim his share for marriage, Hasan felt the weight of obligations towards his parents and other members of the unit and decided to wait until his request came to the top of the list of priorities. The fact that another brother, a stonemason, was able to get engaged that year only emphasized how being part of a large unit could frustrate an individual's personal life plans. After ten years of hard work, he protested, 'I cannot even afford to buy a pair of trousers!'

Hasan was not exaggerating. The shepherds were not given cash, for which they had no need in the highlands: food, clothes, shoes, and their basic needs were sent to them by their mother or sister-in-law, who kept a certain amount of money (either given to them by their husband/father or through the sale of dairy products) for household consumption. During their breaks spent in the town, they were given pocket money. While aware of their assets, the shepherds had no effective power to access them when they wanted to. It was the eldest males who actually distributed the profits and the cash and determined priorities in expenditure. Even younger married men were often victims of this hierarchy. Hasan's own cousin, also a permanent herder in the unit, was still living in his father's tent-house even after his wife had given birth to a boy. His wife was pressing him to split off his share of the herd, sell it, and go back to the town where they could make a new start away from herding. But his plans, too, were discounted because they conflicted with the collective interests of the unit. While marriage gives a couple a stronger reason to establish their independence, the moral pressures of prioritizing the unit and respecting the elders may render them helpless.

As for women, especially unmarried ones, the herd brought them no direct profit. Contrary to the acknowledged Muslim *Sharī'a* law that grants women half the share of men, women did not have shares in the herd. This is different from other pastoral societies in the region, in which women may own livestock. Ownership and the type of property (land or movable) not only determines women's relationship with their natal families but to a large extent establishes their status, whether 'legal' or 'political' (Peters 1980; Maher 1974; Abu Rabia 1994). In Arsal, women were required to contribute their labour for 'nothing in return,' as the unmarried women often complained. The rationale for not granting them a share was that ultimately they would marry and move to their husbands' families. If they did not marry, the assumption was that a male family member (father or brother) would always take care of them. Like men, women gained more control with age and experience, as they moved along their life trajectories. The eldest women assigned tasks and made everyday decisions. Nadia and her unmarried sister were thus expected to obey their parents and offer their labour as part of their domestic duties and kinship obligations. At the end of one agricultural season, Nadia's father was counting the money from a good sale of cherries. In a joking tone, but with serious intent, Nadia and her sister asked for their 'shares.' Their father and mother started laughing and teasing them. 'You make us work and suffer, from cherries to milk to sheep to cleaning, you must give us our right (*haqq-na*),' argued Laila, Nadia's feisty younger sister. Their father finally gave them each a 100,000

LL note (about 60 USD) and promised to buy both of them a golden necklace, which he later did, in appreciation of their efforts.

Married women have a potentially stronger degree of influence because they claim property through their sons, supposing they have them. But if they are childless, like Nadia's sister-in-law for example, the fruits of their labour are more ambiguous. 'My *ta'ab* goes for nothing,' Suad lamented once. 'If I were doing all of this for a *walad* (son),' she continued, 'then it would have been meaningful and worth it. I have treated the shepherds like they were my own [sons]. But now they are old and it is time they had their own wives to look after them.' Unlike Nadia and the other younger women in the unit, Suad saw herself as *bint al barriya* (daughter of the wilderness), thus defying the assumed preference for urban and sedentary living. She spent all her life in the highlands and preferred it to the town. 'The town is so oppressive! People live on top of each other, and the smells ... every time I think of town toilets!' Suad gave a grimace of disgust as she recalled the odours of overflowing septic tanks, a growing problem in crowded post-war Arsal. Her 'toilet' was almost always outdoors, usually in a ditch far away from where the pastoralists slept and where the elements took care of human excretion. For her, modern toilets were unhygienic and repulsive. 'I would never trade the *jurd* for the town,' she reiterated. Her preference was not only related to notions of health and hygiene. Suad was critical of town women whom she thought spent their days in idleness. 'I would die if I just sat. Humans need to do something, they need to move.' The meanings ascribed to 'sitting' will be taken up in the next section. As I will show, sitting is associated with notions of comfort that contrast with the 'running around' that pastoralists do.

While Suad did not want to quit her pastoralist life, she felt overwhelmed by the labour requirements in a unit that managed more than 1000 animals. Without the prospect of having children, the only way to make her life easier was to split off her husband's share to reduce the amount of work. Although women may not have the final say in decisions related to dividing or splitting off from a partnership, they do play a role in influencing their husbands. Large herding units have been known to break up because the wives did not get along, which made daily cooperation disagreeable. Indeed, the social dimension of the herding unit is just as important as the organizational aspect for its sustenance. Decreased unit size implies both social and practical manageability. It means smaller herds and fewer people. But Suad's wishes would still impact her husband's and her own parents' livelihood (since they were cousins and their parents jointly owned the herds). She, too, eventually bowed to the wellbeing of the larger unit at the expense of her own as I discovered when, in August, she began to suffer pain from a tooth infection. Because work is so demanding during that month, she could not find the time to make the trip to the dentist in town, not even for a morning. The drive would have taken her about 45 minutes. So she had to endure another three weeks of pain until the workload lessened before she felt she could get away for the day. At one level, this incident highlights the centrality of women in the agropastoralist labour regime. Suad was indispensable, and this often enhanced her sense of self-worth. But at the same time it underlines the extent to which different pastoralists are bound to a group of others (humans and animals).

As everyone knew, if the system were to survive, *ta'āwun* (cooperation) had to take place between all the members, of different genders and ages. But while this recognition valued men and women in terms of their positions in the constellation of activities, it left their mobility and independence from the unit at the mercy of larger lists of priorities. Projects like marriage could threaten the viability of a herding unit from the perspective of assets, in case men wanted to split off their share, and in terms of labour, if men or women wanted to move out of the unit, as indicated by the individual stories of Nadia, her brother Hasan, and their cousin. It may be that in previous generations marriages did not pose a threat to the structure of households because pastoralists encouraged exchange marriages.[4] Although Nadia's parents would have liked to 'give' (and thus take) more than one of their daughters in exchange, none of the younger women desired this arrangement. Love marriages were becoming the norm in Arsal and young people were pushing to choose their own partners (see next chapter). Often, they preferred spouses who were not herders at all.

The evident growing tensions between different members with diverse interests are perhaps inherent in household relations at large. As others have concluded, 'Household groups are compromises, always imperfect, between often conflicting functional imperatives, pre-existing structures, social norms and cultural standards' (Wilk & Netting 1984: 6). My ethnographic material highlights how some of these negotiations are particular to the emergent changes in the town that are increasingly creating conditions of uncertainty for agropastoralism. The political context of the Lebanese/Syrian border and the ecological changes related to the complexities of land use since the outbreak of the 1975–1990 war have had repercussions on pastoralist organization and movement. These challenges are exacerbated by the rise of new values

---

4   Nadia's mother married her father in exchange for his sister who married her brother. Nadia's eldest brother married Suad (his father's brother's daughter), who was still part of the herding unit. Her other brother, the shepherd Hasan, wanted to marry Suad's sister, his cousin from his own unit. Two women in that case will have moved from one household to the other without leaving the herding unit.

and aspirations that translate into intergenerational tensions. 'Children divide but parents unite,' Nadia's father told me after a heated argument about where best to rent that winter. Dismayed with the options presented, his son Hasan had once again threatened to quit. Perhaps the father was far from oblivious to the prospects of an easier life available to his children who were contemplating alternative lives.

5 Envying 'the Comfortable Woman'

On a cold November evening, Nadia, her sister Laila, and I stayed up late sipping *mate* and cracking pumpkin seeds as we ignored the competing sound of loud snoring coming from the other room where their mother and father slept. The two women were relishing Nadia's return from the highlands to the town. Nadia tormented her sister by saying that this was a 'celebratory *mate*' to mark Nadia's 'last year of service' as a pastoralist. Nadia was hoping, finally, to marry Mahmud by the spring before she could be called upon to join the permanent herders once again. 'Yee!' cried Laila, 'you will leave me to the *shaqāʿ* (exhaustion) as you embark on your life of *rāha* (comfort)!' This contrast between exhaustion and comfort was regularly invoked to describe the difference between pastoralist (whether in the present or the past) and non-pastoralist lives. At the heart of these contradictions was the construction of a new role and status for women. These imaginings were about being modern in ways that were distinct from 'the old ways.' Lara Deeb suggests we explore 'not only local understandings of being modern but also how these understandings are employed and deployed in various contexts and to what effects' (2006: 15). In the context of the uncertainties facing agropastoralism in Arsal, these local concepts of modern-ness contrasted with the pastoralist way of life and rendered such lives non-modern and undesirable.

In discussing the ideal they aspired to, the two sisters brought up their friend Ghinwa, whose life as a *murtāha* (comfortable) woman was the object of envy. Ghinwa was married to a truck owner who was considered relatively well off, earning an average of about 1000 USD a month in 2003. This was reflected in her lifestyle, the modern goods in their house, and in the money Ghinwa spent on herself and her family. Her comfort was evident in the amount of leisure time she could afford as a young housewife. Everyday Ghinwa would find a ride to her parents' house after her husband left in the morning. She would spend hours with her sisters or friends from her old neighbourhood. After lunch, she would shop in a nearby grocery store and return to clean the house and prepare an evening meal for her husband. 'She sits in her house with her two children, she buys whatever she needs from the shop. No highlands, no sheep, no sun, no hard work,' Laila explained. She then asked us if we had noticed Ghinwa's 'beautiful pale skin.' Burnt tanned faces indicated hard work in the fields. Pastoralists of all ages tend to have chapped complexions and freckled faces because of the amount of time they spend in the sun – an increasingly undesirable standard of beauty for women who preferred to have pale, spotless skin. The idea of comfort seemed to lie in the fact that Ghinwa was 'cared for' (*ya'tanī fī-ha*), evident in her daily routines, consumption practices, and bodily presentation and comportment.

My interlocutors were echoing a desire for an idealized form of domesticity. Laila's use of a mobility idiom, 'sitting,' was a common way of articulating the specialized and limited role of the non-pastoralist housewife/mother who is *qā'da li bayt-ha wa awlād-ha*, literally 'sitting for [tending] her house and children.' The stillness in 'sitting' is significant and conveys a degree of luxury or privilege in the description of the *murtāha* woman, unlike pastoralist women who describe their lives as 'constant running' (*rakd bi rakd*) that ultimately leads to exhaustion. These are expressions of pace – the slower the better – that unfold in household types that have developed with the new livelihoods outside of pastoralism. This type of domesticity was made possible by the emerging preference for neolocal residence that signalled the autonomy of a newly established family unit. By the early 2000s, living with a husband's family was no longer desirable and was even considered to undermine women, so much so that engagements lasted for very long periods until the groom was able to build a new house for his bride. A separate, independent house was a condition across the board, even for women who practised agropastoralism and lived in the highlands for the major part of the year, as we saw earlier in the case of the shepherd who wanted to marry his cousin. I knew couples who had been waiting for three years, sometimes five, for the completion of their house. Given the nature of seasonal agricultural work and day-wage labour in the quarries, there was a general understanding that house construction was contingent on interrupted income and availability of cash. Parents, however, tended to refuse proposals from suitors who showed no intention of moving their wives to accommodation that reflected independence from the man's natal family. Although houses were not given to women, in the sense that they were not registered in the wife's name upon marriage, as might be the case in cross-cultural dowry practices, their construction was seen to honour the autonomy of the couple as a new family and economic unit, and to recognize the 'reign' of the new wife over what people termed 'her new kingdom' (*mamlakah*).

The conception of household as *mamlakah* offers an interesting insight into the transformations of gender and kinship relations. The image of the 'reign' of a king and queen in the house derives from the changing kinship relations consequent upon the removal of labour from the realm of the family. In *mamlakah*, the male guardian (the husband/father) is designated as the main income provider while the wife autonomously rules over the realm of homemaking, without having to submit to other women and matriarchs. This is a significant contrast with pastoralist households whose members, as this chapter shows, are tied to the economic agenda of the joint and extended herding unit. The change in domestic governance (Mundy 1995) is not entirely about the formal lines of division of labour, but rather points to transformation in content and the sharpening of the boundaries of men's and women's worlds. In the agropastoral regime, women are burdened by the intensive year-round labour that constitutes an essential part of their domestic duties. While most work is gendered, the intensity of labour demands at certain times of the year and the predominance of the ethos of cooperation mean that the 'domestic' is often elastic and blurred. The guidelines of decision-making and responsibilities described in previous sections were not inflexible. Women were involved in decisions about camp location, the sale of sheep, and even men's work such as shepherding, particularly during times of shortage when these matters were passionately debated in the family. Similarly, there were instances, albeit fewer, when men participated in 'female' work and decision-making, for example milking. Therefore, where a division of labour did exist, the boundaries were pliable and men and women could easily cross them when needed. Although (older) men were in control of the larger, more vital resources of the herding unit, agropastoralism, as a regime of production, placed women at its core.

In the *mamlakah* model, women were also central to the household; however, the boundaries were becoming sharper with the move away from family labour. As men turned to wage labour, female self-making was increasingly tied to homemaking. In this new domesticity, women's roles turned inward to their house and household, now that agricultural activities no longer constituted the domestic. Marginalized from the process of production, women who were not in formal employment were reduced to their reproductive roles.[5] Though not all pastoralists desired this female ideal (recall Suad who looked down on the idleness she saw in this model), it was increasingly gaining traction. Even

---

5 Wilk and Netting write that reproduction includes activities related to child bearing and raising: 'Reproduction is a stage in a process whereby time and effort are invested in offspring who may later increase the household labour pool and tend the aged parents when they become physically dependent' (1984: 14). Later, Caroline Moser (1993) made a similar argument.

for formally employed women who worked as teachers or in NGOs, homemaking remained the arena in which female personhood was formed and performed. Men generally expected their fiancés to leave their jobs after marriage. Those who encouraged their wives to continue working in certain paid jobs such as teaching demanded that these activities did not conflict with a woman's perceived domestic duties: keeping a tidy and clean house, ensuring the husband and children's apparel was clean and presentable, cooking and nurturing, managing household budgets, and demonstrating an ability to entertain were all part and parcel of being *rabbat al-manzil* (mistress of the house). While employment and education enhanced a woman's sense of self, they sometimes conflicted with this model of domesticity. I was told that men did not like to have a more educated 'superior' wife, though having some education enriched the role of the mother who was expected to help her children with their schooling. Education was therefore seen as a vehicle for the creation of the modern housewife whose life ought to centre around her husband and children.

If households are understood as 'the structure of activities that lead kin to stay together' (Mundy 1995: 100; Guyer 1981; Netting et al. 1984), then placing these activities and the relationships surrounding them in the context of wider political and economic structures allows us to trace ethnographically how gender and kinship are reconfigured at a time of social transformation. As the material in this chapter suggests, households are also units of cultural meaning (Yanagisako 1984) that gauge emotions and sentiments. The economically independent, neolocal, nuclear household is an aspiration articulated in the notion of the *mamlaka*, reigned over by a 'comfortable' woman who is 'looked after' by a husband. It is these conditions that are seen to allow women to specialize in their reproductive duties and to limit their scope of domesticity to their children, husbands, and their town house. The expectation that a bride would walk into her own *mamlaka* (kingdom) is very much a function of interlinkages between socio-economic transformations, changing gender roles, and new trends of consumption in the area. 'Why live in a tent-house, move at least three times a year, spend your life tending to animals, fetching water and suffering *shaqā'* (exhaustion) if you can stay in a town house, equipped with everything you need?' Laila wondered. An understanding of households in the Arsali context, however, is incomplete without an appreciation of how evolving gender roles and the emerging new domesticity of women was (re)shaping what constituted the normative expectations and rights in marriage. The next chapter explores transformations in Arsali marriage.

CHAPTER 5

Marriage between Love and Fate

On our way to the highlands, a telling conversation took place as Fuad recounted to Hana and me the occasion of his daughter's engagement, which had taken place the week before. Fuad considered himself '*munfatih*' (open-minded), for he had noticed the frequent visits his nephew (his sister's son) was paying his daughter and concluded that 'something was going on,' yet he had not interfered in this developing relationship. An advocate of romantic love, he approved the match as soon as his nephew approached him formally. Our discussion moved on to kin marriages in general, and I took the opportunity to ask about the traditional father's brother's son (*ibn al-ʿamm*) marriage, which remained a common, if not preferred, marriage in Arsal and other parts of Lebanon. 'Those marriages were more predominant as we go back in time,' Fuad replied. Explaining why he thought people preferred them, he added that 'the preference for *ibn al-ʿamm* marriage is based on the fact that two brothers provide the *āmil zukūri* (male factor) which is *aqwa* (stronger).' When I asked him to elaborate on this notion of 'patri-strength,' Fuad paused to consider, as if he had never really given this matter much thought before, and ended up with the following: 'I suppose stronger from a reproductive point of view (*injāb*).' Being a supporter of this form of marriage, Hana changed the direction of the analysis and stepped in with a more confident justification. '*Ibn al-ʿamm* confers the obligation to protect his cousin.' Fuad, who did not seem to care about 'old-fashioned' marriages, tried to bring the discussion to an end. 'These days kin marriages no longer take place. In fact people now prefer external marriages.' In view of his daughter's recent engagement, I decided to challenge him: 'But doesn't your daughter want to marry her cousin?' 'Yes, but it was her own choice (*khayār*). The two of them *want* each other (*bi-rīdū baʿdhum*),' he replied. I continued my provocative line of enquiry with a hypothetical question: who would Fuad prefer if Rana had had two equally good suitors, one from his family and one from another lineage? His answer was conclusive. 'The family, for sure! The family provides grounds for protection (*himāya*). No other institution such as the state or any other fulfils that role.'

Although liberal-minded people like Fuad may denounce cousin and kin marriages as 'old-fashioned' remnants of the past, the tradition seems to persist not only as a practice – chosen by Fuad's own daughter among many other young couples in Arsal – but also as a powerful cultural ideal that merits investigation.[1] What are the aspects of kin marriages that continue to appeal to contemporary Arsalis? And what precisely are the aspects that people like Fuad dismiss? The brief conversation above points to two key issues that require some attention in the analysis of Arsali marriage. The first is consent, an idea that is often expressed in the language of 'wanting' (*irādah*) and 'choice' (*khayār*), and is perceived by Arsalis to exemplify the way marriage is undergoing transformation from the not so distant past. Nowhere can the concept of 'temporal consciousness' (Limbert 2010) be more resonant than in the area of marriage. My older interlocutors were keen to emphasize the extent to which practices and values surrounding marriage have changed, sometimes reminding younger people of their privileges 'today' compared with older generations who often had no say in decisions about their future. The second issue is the currency of previous kinship in a marriage and its perceived capacity to provide what Fuad referred to as 'protection' (*himāya*) among the other advantages that ensure a successful relationship. What are the threats to marriage in contemporary Arsal that require *himāya*?

It is possible to observe some common cross-cultural trends governing change in contemporary marriage. Shifts in market economies, for instance, are particular drivers of change in marital and kin relations. Arsal is no different in this regard, given the decline of agropastoralism. The scholarly discourse on marriage identifies one particularly noticeable change: the waning centrality of kin in arranging marriage, now replaced by expressions of 'individual desire' (Collier 1997; Gullestad 1996; Sabean 1990; Yan 2005; Hart 2007; Ahearn 2001; Giddens 1990). This shift is often articulated in the vocabulary of 'modernity' that upholds the rhetoric of romantic love, coupled with 'individual choice' and 'autonomy' (Hirsch and Wardlow 2006).[2] The main theoretical question in the ethnographic account of marriage seems to revolve

---

1 My sense is that general doubt about these marriages, stemming from widespread understanding of their genetic implications, is increasing in more urbanized areas of Lebanon. Although there are few contemporary studies on consanguineous marriages in Lebanon, one recent work confirms previous findings that suggest 'consanguinity frequency' in Beirut is 28.6% (Barbour and Salameh 2008: 7), compared to earlier estimations of 25–26% (Khlat 1988; Khlat and Khudr 1984).

2 For example, Jane Collier's work with Spanish villagers highlights the contrast they made between 'letting others think for you' and 'thinking for yourself' (1997: 5). Similarly, in Norway, Marianne Gullestad (1996) identifies a parallel shift from 'obedience' to 'negotiation' in family values that seem to emphasize 'autonomy' and 'individualism,' so characteristic of 'late modernity.' Yunxiang Yan also shows how recent economic shifts and 'westernisation' in China have led to a rise in autonomy in marriage choice as well as transactions surrounding marriage (2005: 639).

around 'how individualism emerges as a location for romantic love, personal expressions of desire, and whether this necessitates a split from extended kin networks' (Hart 2007: 346). David Lipset suggests that 'romance may become a tactic in an intergenerational politics (asserted by youth) for independence from parents ... and in a cross-gender politics (asserted by women) for power vis-à-vis patriarchy' (2004: 208).

My material echoes some of this literature, particularly with regards to the perceived changes in intergenerational negotiations surrounding marriage, where parents no longer feel they can (or should) impose the same kind of authority exerted by earlier generations. In many ways, this has to do with changes in gender roles, particularly women's, in affecting how people think about marriage. In keeping with some of the maxims cited above, young Arsalis also expressed the desire to 'live one's own life' (*yaʿīsh hayātuh*) rather than 'lives dictated by others.' Yet, where there is a call for a level of autonomy in deciding on one's future spouse, the question of 'individual desire' is a matter that needs unpacking as it implies neither a split from kin networks nor a disavowal of parental authority and approval. One of the problematics of the concept of individualism is that it invites dichotomous theorizing that assumes that individualism precludes and excludes forms of 'sociocentrism' (Kesserow 1999). To avoid this, I argue that we need to situate this language within social and normative matrices that dictate the nature and limits of desire in a particular place (Joseph 2005). I take on board Kimberley Hart's recommendation to investigate marriage as 'the intersection of intimacy, economic and kin ties, evolving gender roles, and transforming cultural practices on local, national and global levels' (2007: 346). There is perhaps one more element in the Lebanese Arsali case that complicates these inter-linkages, and this is the conceptualization of fate (*nasīb*) as a determinant of people's lives.

1   The Befalling of Nasīb

In Arsal, as in various parts of the Arab-speaking and Muslim world (c.f. Ahearn 2001), fate (*nasīb*) features heavily in how people understand their lives. *Nasīb* is believed to predetermine life. One's fate, I was told, is 'written on one's forehead' by God who ultimately decides the fate of every individual even before they are born. Although humans cannot 'see' the writing, believers know that it exists and contend that it was 'inscribed' upon their birth. Whatever happens to them later is thus 'from God' (*min Allah*), an expression that sums up an attitude we have come across in previous chapters. *Nasīb* is used to justify all sorts of events such as divorce, death, illness, bad luck, good luck, and livelihoods too. But aside from determining life in general, *nasīb* has a special association with marriage and is used as its metonym. Like fate, marriage is characterized as a mystery for it constitutes the unknown and is revealed only after it takes place. Other idioms of marriage articulate similar ideas. For example, a common saying is that 'marriage is like a watermelon (*battīkha*), we cannot know its inside until we break it open.' The serendipitous aspects of marriage are articulated through the common expression 'the fate has befallen' (*ija al nasīb*).

Fate as the Arsalis imagine and talk about it, as a metonym for marriage, is gendered. Its relationship with women is testing, impatient, and short-tempered in a way that it is not with men. Marriage fatalism applies to both genders, but it seems to put women under particular kinds of pressure. Although it is common for women to marry in their mid-twenties, and sometimes even later, the age of marriageability expires much sooner for women than for men, who are able to find a wife no matter how old they are. And certainly older widowers, even in their late seventies, are encouraged to search for younger wives to look after them. Men can also take up to four wives, at least according to Islamic Sharī'a. However, in spite of it being religiously sanctioned, I found that polygamy was a controversial topic, and that men very rarely married more than two wives. My interlocutors generally doubted that men could treat women equally, in accordance with the stipulations of polygamous marriage in Islam. Men who married a second wife without good reason were ridiculed for their excessive sexual appetite. One of the main acceptable justifications for taking a second wife for men is the inability of a first wife to conceive.[3] Nevertheless, the potential occurrence of polygamy, given that it is not against religion, gave women cause for worry. I often heard men tease their wives about taking a new wife. Some wives ridiculed these threats and saw them for what they were – 'tools of manipulation to keep us on our toes,' as one woman told me, shrugging off her husband's banter. Others felt less confident.

For women, the stakes were much higher in terms of when and if *nasīb* befell them. In Arsal, womanhood was attained through marriage, until when women were categorized as girls (*banāt*), regardless of their age. In her analysis of Palestinian womanhood, Amalia Sa'ar suggests that this 'linguistic gesture of infantilizing and desexualizing unmarried women represents a normative

---

3   I have known couples who have been unable to conceive because of the man's medical, or other, condition. In such cases, the marital couple may remain childless. When a woman cannot conceive, however, the husband is encouraged to take another wife, sometimes with the approval of the first. Also, it could be that the man cannot conceive but takes a second wife anyway, thinking that the problem lies with the first. With increased access to medical treatment, many people resort to available, religiously-sanctioned new reproductive technologies (see Inhorn 2007, Clarke 2009).

expectation that the passage to womanhood should occur in a specific, institutionalized form' (2004:1). Implicit in the term is an assumption about women's lack of sexual activity and the preservation (*hifāz*) of a virginal state. In a setting where fulfilling womanhood and attaining an adult form of femininity rested on marriage, it is unsurprising that women experienced high anxiety in the period between reaching marriageable age and passing it. The anguish was related to becoming a spinster (*'ānis*) and to being stigmatized as a failed woman, a grown-up 'girl' who never achieved womanhood. This is not to suggest that unmarried women did not occupy a central role within their families and communities. The varying roles and positions of women, their sexual ventures in spite of the expectations and codes, and their general independence, have all been documented in the literature (see Sa'ar 2004; Baxter 2007), and Arsal is no different. Yet the cultural ideal of womanhood was threatened in the event of an individual remaining single, not to mention the distress over the reconfiguration of authority in a household once the parents were deceased.

Single women never lived by themselves, unlike some of their urban counterparts, and ended up being the responsibility of a male kin, sometimes a younger brother. In the next chapter, I explore in more depth the common fear of 'enslavement' by the brother and his wife, who exploit the burdensome single sister in return for keeping her in the house. It is precisely during this liminal period that women were reduced to waiting subjects (Elliot 2016). Here the notion of fate can be understood in its temporality, with the relentless waiting, pregnant with hope and anticipation eventually descending into frustration (Gasparini 1995). This liminal state is exaggerated in extended families that favour cousin marriage. It used to be a common practice for parents to wait until all a daughter's paternal male cousins were married before they would consider 'giving her away' to other potential suitors, especially non-kin. Jamila was one such woman who was under the impression that she was 'reserved' for her older paternal cousin. When he married someone else, her parents expected that one of his brothers would ask for her hand. There was never an open conversation about this matter between her father and uncle. Her father, she told me, was too proud to ask his brother, so he assumed that they had an understanding. By the time her male cousins were all married, she was already categorized as a *'ānis*, too old to marry. She felt left behind and heart-broken. Perhaps this is the sentiment being conveyed in 'waiting for *nasīb*,' an expression used mainly of single women.

During my fieldwork in Arsal, I witnessed many a woman's emotional angst on reaching the cusp of marriageability. Faced with the tyranny of time, choice seemed to fold in on itself as the options narrowed. From the teen-age romantic dream of awaiting the *habīb* (lover)-turned-husband, older single women either accepted a life of spinsterhood or settled for an otherwise undesirable husband: an elderly man, an already married man, a poorer man, etc. I watched my friend Maha agonize over accepting a suitor, just when she had begun to lose hope. Aged 30, she was second of seven siblings. Her parents were struggling to provide for their children and give them a good life. Maha began to feel that she was adding to her parents' burdens, especially when one of her younger brothers fell terminally ill and the family could no longer cope with the exorbitant costs of Lebanon's private medical care. The prospect of marrying a suitable man from their social and kin circles had dimmed. Out of the blue, a herder she had never met approached her parents. His family had a good reputation but they lived in a herding unit, in tents in the lowlands on the border of Syria. Maha fretted over her decision for days. She was scared that her life was going to deteriorate, and that the wilderness, the labour, the sheep, the distance from the comforts of the town would take their toll on her. She was losing sleep over allowing herself to walk into a lifestyle that her peers were escaping. But rejecting this chance was a big risk, as she had no guarantee against the prospect of spinsterhood looming over her now that she was 30. She finally agreed to the marriage. Concerned that her decision was made out of sheer desperation, I asked her if she was convinced (*muqtani'a*). Her reply was as fatalistic as any could be in the Arsali marriage landscape. 'Don't they say *nasīb*? Mine has come and I must take it.'

The linguistic usage of *nasīb* simulates a lack of agency by the individual: marriage befalls, it arrives and it happens (or not), even when the individual herself makes a decision, like my friend Maha. This expression of docility in the face of marriage is arguably pertinent to a setting where, in the very recent past, it was not unheard of to marry a complete stranger. The extent to which marriage is perceived to have changed is nowhere clearer than in narratives that contrast contemporary practices with *al-qadīm*, 'the old days.' Central to these narratives is the powerlessness that women (though sometimes men too) possessed against obstinate parents or guardians whose authority was unquestionable, sometimes to the extent of their exercising violence on those who disobeyed their authority. One woman expressed disgust with her grandfather who had apparently tied his daughter to a pole and whipped her because she was in love with a man he disapproved of.

Another woman, Khadija, was forced to marry her maternal cousin around the age of eleven, before she had even started her periods. Initially, her cousin was engaged to another woman. But on the first night of the conventional three-day wedding party, the bride's family called the marriage off because they had decided to 'give' the bride to her 'rightful' paternal cousin instead. In a moment of extreme tension, Khadija's elder brother turned to the groom's father and, in front of an expectant audience, dramatically offered Khadija, who was skipping with a rope outside, as the substitute bride. In shock, some protested that 'Khadija is still a child!' But Khadija's brother had already made his public pronouncement and was not going to retract. It was the time of year when Khadija's father and sisters were in Syria with the rest of the herding unit. Her elder brother did not want any delays. Khadija told me that when he informed her mother and herself that the groom, his family, and the *shaykh* were on their way for *katb al-kitāb* (the Islamic marriage contract), her mother started yelling. And so did Khadija,

> I started screaming until they heard me in Beirut. Then they started instructing me to answer [the *shaykh's* questions]: 'Who is your *wakīl* (custodian)?' I was a child, only ten or twelve! I hadn't even started menstruating. My periods came two years after I married. Anyway, they wrote the contract and did what they wanted. For three days, I shut myself in the room and cried. But then, the *nasīb* took its course.

People eventually yielded because there was a serious issue at stake: the honour and face of the lineage, as Khadija explained, sarcastically stressing the Arabic word for honour, *sharaf*, as if critiquing its cost. Khadija was the solution 'from within,' against her will. Her fate landed her in a long and unhappy marriage. Her husband was much older. Even in his late 60s, she complained, he was still a difficult and jealous man. 'We spent our lives fighting! From the first day we fought. Bitter is the fate that has befallen me.'

There is general agreement among the older generation that a couple's approval was not a prerequisite for marriage. Rather, my elderly interlocutors accepted the preference for exchange and kin marriages. Embedded in Arsal's economy, marriage practices organized households around the sustainability and needs of their livelihoods. This included keeping a balance between labour and ownership of property – namely land and herds. However, these preferences ought not be seen through their functionalist lens alone but rather as an ideology of proximity, as I will show in more detail in the rest of the chapter. This ideology thrives in certain economic regimes. My interlocutors spoke of 'departing' when they described women's marriages: 'Women *b-titlaʿ* (col. depart) from their natal families and join (start to belong to) their husbands.' This shift does not necessarily impinge on everyday relations and closeness with natal kin. But there is always the *fear* that it might, especially among agropastoralists, when a daughter's herding unit might be camped as far away as the borders of Iraq. In this rationale, 'departing' as close as possible – to a father's brother's house – guarantees a daughter's proximity as well as her committed labour. In this context, cousins would be 'assigned' to each other from childhood and parental authority over this matter was normalized.

Romantic love and consent were not basic criteria for marriage; several interlocutors described to me their sense of drifting into lives dictated by others. The bitterness expressed by Khadija, however, was not always an inevitable outcome, for several people ended up with happy marriages in spite of not having consented at first. But it was precisely extreme narratives like Khadija's that were invoked to reflect on some of the changes that had taken place in Arsali social life over the decades. The use of coercion in contemporary marriage was highly disapproved of. Increasingly, education, literacy, and the spread of satellite television contributed to a broader knowledge, not least around what constitutes 'sound religion.' Marriage without consent was not just morally and socially contested and legally prohibited, but also religiously forbidden (*harām*). Some might argue that marriage decisions were reached through negotiation, possibly even arm-twisting and emotional blackmail. But using force, let alone with someone still thought to be a child, was no longer tolerated and was seen to be uncivilized, even savage. To Arsalis, stories like Khadija's had become a reference to practices that reflected a time gone by, when one's desires and rights were irrelevant and when fates were driven by seemingly rigid patriarchal structures.

2   The Vocabulary of Modern Marriage

The right to establish an independent life outside of paternal families has become a legitimate one for conjugal couples in post-war Arsal. There was a noticeable change towards contemporary marriages where, as Fuad emphasized at the beginning of the chapter, young men and women followed their desires and exercised their *khayār* (choice). But the choices made by couples, even in these emerging expectations, were not without negotiation that sought parental approval and worked around consensual family politics. In this section, I discuss the vocabulary of modernity deployed in these negotiations, particularly the rising rhetoric of romantic love as a necessary requirement for a happy marriage. Anthropologists of emotion have strongly contested the Eurocentric argument that love is a novelty outside of western societies and that its surge ought to be understood as a function of modernity. But David Lipset contends that outside of Europe, love discourse (rather than the emotion) *is* perhaps a modern one (2004: 208). Laura Ahearn also shows how 'desire itself came to be seen as desirable' in Nepal (2003: 107). It can fairly be argued that transformations in love and intimacy are intertwined with other forms of expression – political, economic, and ideological – that reflect wider social processes (Hart 2007; Toren 1999).

2.1   *Love and Desire*

To admit openly that one is in love remains a contentious issue in Arsal, often bordering on shame (*'ayb*). There are various expressions for love. *Hubb / mahabbah* is a general sentiment of affection and is often used to describe feelings for children or among them. *Mahabba* especially is used to express general caring and warmth among people and their communities. *'Ishq*, on the other hand, is a much more intense emotion that is sexually charged and dangerous, for reasons similar to those outlined by scholars of love in the Middle East (Hart 2007; Marsden 2007a). Lila Abu-Lughod's (1986) work on the Bedouins of Awlad Ali articulates the threat that love poses to ideal family control over marriage in a kin-based society as love leads people to lose self-control and thus threatens social obligations and ultimately the social order. This rationale is expressed in comparable tones by the older generation in Arsal. People who fall in the *'ishq* type of love are ridiculed if they act outside the expected norm. For example, Zuhayr and his wife were very concerned about the behaviour of their daughter's fiancé. While visiting one's fiancé is common practice, it has to be done within reason and decorum. Their future son-in-law, however, was in their house every day, for long hours, clinging to their daughter. This caused alarm as Zuhayr expected him to be working hard to finish building the marital house rather than staring adoringly at his daughter day and night. The fiancé's behaviour was embarrassing. 'This is inappropriate!' Zuhayr protested scornfully. 'You'd think someone would snatch her if he removes his eyes from her. He isn't *'ashqān* (in love), is he?' His comment suggested that he had lost his mind and social aptitude. In the highlands, the herders made a joke of their bellowing donkey. It was '*'ashqān*,' they would say, and only that could explain the animal's loud painful desperate cries that were inhibited by neither the public nor its judgement.

Although public displays of *'ishq* are criticized in everyday practice, *'ishq* is paradoxically celebrated in the realm of popular culture. Classical genres of folklore, especially song, take *'ishq* as one of their central themes. This includes *ghazal*, improvised songs that are based on rhymes (see Marsden 2005, 2007a), *'atābah*, a genre of song that laments lost love, and *dabka*, a more lively genre associated with a step dance, often used at weddings. In all these, romantic love is idolized and lovers tell stories of waging wars for their loved ones, defying parents, challenging a female lover's unbending male kin, climbing mountains, and crossing seas. The powerful rhetoric in traditional genres attests to the recognition of love among the older generation, despite the need for it to be regulated in Arsali society. Romantic love is increasingly being popularized and consumed in contemporary forms accessible to Arsalis through radio, magazines, satellite television, and more recently the Internet. Long evenings at home were spent watching dubbed Brazilian and Turkish soap operas and Syrian dramas, all of which revolved around intense love stories. Even in the highlands, when labour was most demanding, the treat at the end of a day's hard work was '*tamthiliyyat 'Arab*,' a Bedouin soap opera aired on Syrian TV that we watched on a tiny black and white set powered by a car battery. We followed the story of a Bedouin man separated by evil forces from his father's brother's daughter, who faces wild desert adventures to get her back.

The discourse of romantic love was gradually gaining currency at the time of my research. Both *hubb* and *'ishq* were especially prevalent among the young. I listened to teenage girls verbalize daydreams of love. One, for example, told her cousin and me how she would give her whole life to her fantasy future husband. This was expressed through the language of new domesticity and homemaking: 'I will make everything perfect for him. His shirts would be as white as snow, ironed to the millimetre [perfectly]. I will even iron his socks! I will prepare the most beautiful food for him so when he comes home he will be pleased.' I also heard engaged women joke (sometimes crudely) with their friends about sexual intercourse on the waited-for wedding night, considering this was meant to be the first time. But neither lust nor love was articulated as a rationale for marriage. These cheerful conversations and daydreams had their place in particular private social settings and did not qualify as a coherent foundation on which to build a *zawaj sālih* ['good or sound marriage']. Citing 'love' as a reason for marriage was still seen as inadequate, if not inappropriate, and something that resonated more with TV drama rather than real life.

If elders sympathized with romantic love, they deemed it insufficient in intergenerational marriage negotiations. This rhetoric of romantic love is complicated by expressions of desire. The idiom of *irādah* incorporates 'wanting' (the verb to which is '*yurīd*,' to want) as well as having a 'will.' Someone has a strong *irādah* when they are determined to do something (to complete a challenging task) or even to resist something (to abstain from certain foods). The use of this expression in both its connected senses of wanting and willing is very telling and points to the efforts to 'live one's own life' as opposed to 'lives dictated by others' that the tales and experiences of the older generation evoke. But desire can also be as dangerous as *'ishq* when it is driven by *raghbah*, a bodily appetite that is not always voluntary (like a *raghbah* for sex or food). It is an overpowering feeling that could and ought to be controlled when its irresistibility begins to become overwhelming. All these aspects of desire may well overlap and intertwine. When connected to romantic love, *raghbah* in its

FIGURE 11   Father of a bride dancing on the first night of the wedding

own right, like *'ishq*, is discredited. Desire, therefore, becomes a valid expression and a tool of negotiation only when it is embedded in wider ideals surrounding the meaning of marriage.

2.2   *Partnership and Understanding*

To the Arsalis, marriage is an apparatus for facing life. In a now familiar fatalistic framework, my interlocutors expressed the belief that God gives us our lives but he predetermines how we live them. I was often reminded that life was short and naturally fraught with obstacles and hardships, not least through the frequently quoted mantra: 'Life is but a couple of days that need to pass, whether in goodness or in ugliness' (see Introduction). The inevitability of hardship, however, can be mitigated through the locally perceived pillars of marriage: partnership and understanding. By partnership (*tashāruk*), people denote an ethos of sharing between a heterosexual couple who are expected to have children and create a family, which leads to the reproduction of society. This partnership is not necessarily egalitarian, as men and women have different roles, rights, and obligations. But a partnership makes the journey of life

FIGURE 12 Bride and groom leaving the wedding ceremony to their new home

less challenging, more so if it is based on understanding (*tafāhum*),[4] or the meeting of minds, so life can pass in goodness rather than ugliness. For the older generation, partnership and understanding, if any, unfolded temporally as *nasīb* slowly but surely revealed itself and as a couple lived together over time. Amidst the uncertainty characteristic of life and fate, it becomes necessary to consider who one is likely to have common grounds within ways that will counteract the inevitable imminent hardships.

This question is partially answered by the meaning of marriage expressed linguistically through the term *yata'ahhal*. The expression carries two meanings. The first is 'to qualify,' an indication that the couple is mature and ready to move on to a new life-stage. The second is 'to become *ahl* [kin].' Marriage, then, is about creating or recreating kinship, which, to a large extent involves forging long-lasting if not permanent, irreversible relationships. When marriage constitutes the unknown, it becomes plausible for people to resort to strategies that play down 'threats' (Bourdieu 1966). In a society like Arsal, which naturalizes kinship and sees kin relations as entailing a pre-ordained morality of trust, marrying kin becomes a guarantee for a secure marriage because kin have an element of predictable sameness. In this sense, marrying kin results in what Martha Mundy has labelled the 'redoubling of affinity' (Mundy 1995: 182). This argument makes sense largely within economic regimes, such as agropastoralism, that are based on collective family labour. Living arrangements organized around labour mean that proximity, familiarity, and conceptions of natural sameness among kin guarantee a continuity in, if not a 'redoubling' of, both partnership and understanding. Within this ideology, then, desiring a close kin with whom one already has the necessary proximity and social bonds guarantees a successful marriage. This ideology is upheld by the younger generation as much as the older, except that the young are more emphatic about desire.

In the spectrum of a rapidly growing population where kinship circles seem to expand exponentially, it is possible to be unfamiliar with one's kin. Unless relations of *'ishra* (living together) are nurtured, the assumption of 'natural' sameness based on the kin bond may have to be put to the test. The younger generation have more avenues for meeting people through school, emerging small businesses, public transport, local NGOS, and university than the older had. In this case, they may develop understanding (if not emotions) with

---

4 Nancy Jabbra who worked in a village in the Biqaʿ region has similar findings on how attitudes to marriage in the post-war period were based on a concept of 'understanding'. She cites an informant explaining that 'the relation between husband and wife … should … be based on understanding. As for the relation with their children, it should be based on love, respect, and understanding' (2008: 69).

individuals who are not the right kin or not kin at all. But equally, close kin may be too close for comfort. Too much sharing and understanding, such as the one replicating sibling relations, for instance, may inhibit *raghbah* (desire) and sexual attraction, as we will see in the next section. It is precisely in this elastic space of closeness that 'wanting' someone becomes an issue of negotiation. 'Understanding,' as a pillar of marriage, needs to be subjective, consciously recognized by the individual, not just assumed by parents or other figures of authority. In the remainder of the chapter, I wish to draw out this discourse of wanting in marriage as it emerges ethnographically, particularly in relation to a resilient cultural preference to kin marriages against a growing discourse of desire in a context where both parents and children are working with changing values.

3 Intergenerational Negotiations

Based on her work in Örselli village in Turkey, Kimberly Hart rightly problematizes the trend in some analyses to treat love matches as an index of modernity that favours individualism, where young people challenge parental authority and express individual desires (e.g. Giddens 1990). She found that 'rather than a radical break between the authority of kin and the individualism of youth, both young and old make accommodations to arrangement and romantic expression' (Hart 2007: 347). Other research on Turkey echoes this (Tekçe 2004) and suggests that even in arranged marriages, parents take their children's emotions into account (Duben and Behar 1991). A similar argument can be made for Arsal. In this section I look more closely at Zuhayr and his family, with whom I had the opportunity to live for a period of time. The parents' self-reflexive considerations of marriage shed light on the extreme tensions created through expressions of 'wanting' that departed from the predilection for kin marriage.

3.1 *Zuhayr and Rima*

Zuhayr, the NGO worker we first encountered in Chapter Two, considered himself to be a progressive person. He worked hard for his family and had aspirations to provide his children with a more prosperous life than his own. Particularly, he had high hopes for his two elder daughters. He had encouraged their university education and had even discussed with several university professors the possibility of securing scholarships for them to study for postgraduate degrees in Lebanon or even Europe. Both daughters were already in their first and second years of university and Jumana, the eldest, worked as a local schoolteacher at the same time as studying. Finding a school job in Arsal was not easy. Jumana had applied several times before securing a post in this competitive and small field. As his daughters reached their early and mid-twenties, Zuhayr began to receive marriage proposals from close kin eager to marry their sons to his two good-looking daughters. This growing pressure proved testing to both Zuhayr and his wife Rima. Neither of them wanted to replicate the cruelty of an imposed marriage as they had experienced it. Yet, as their story elaborates, despite years of unhappiness, their marriage was living proof of the resilience of close-kin marriage.

To those who did not know them well, the marriage of this couple was always an ambiguous affair that tested the limits of proximity, blurred the boundaries of social and biological relatedness, and complicated the discourse of the unfair *nasīb*. A main confusion in their relationship was that they shared a brother, Ahmad. I overheard two men on a public bus trying to solve this riddle. How can Ahmad be Jumana's *'amm* (father's brother) and *khāl* (mother's brother) at the same time? Surely Zuhayr and Rima must have committed some forbidden act of *harām*,[5] inevitably a cause for condemnation or repugnance. Those who knew them closely understood that these concerns had perhaps occupied the couple just as much at one time. In fact, Ahmad was Zuhayr's brother from his mother's side and Rima's from her father's.

When Zuhayr's father died, his mother was left with four children to raise. Zuhayr's uncle (father's brother), who was already married with a daughter, Rima, decided to help relieve his late brother's wife[6] of her burdens and took her as a second wife. This led his first wife, Rima's mother, to accuse him of having a voracious appetite for women and to divorce him. Soon after, she married a relative and left Rima with her father and new stepmother. Ahmad was born two years after their marriage.

When he was 20, Zuhayr fell in love with a young woman from the town. But when his uncle found out, he willed him to marry *bint 'ammuh* (his father's brother's daughter) Rima or leave his house for good. This demand left Zuhayr with very mixed feelings. On one hand, he felt he owed it to his uncle to show gratitude for raising him and his brothers. On the other, he and Rima were raised as siblings. 'How can a man marry his sister?' Zuhayr fretted, as he

---

5 The insinuation here was incest. Clarke (2009) reminds us that there is no Arabic equivalent to the English word 'incest' and that, rather, *zinā*, fornication before marriage and adultery after it, might be a more appropriate category (see also Van Gelder 2005).
6 Levirate marriages are not very common in Arsal. Although they are not *harām* (forbidden) religiously, they are not well-regarded. In terms of Zuhayr's story, the repugnance was not due to his mother's remarriage. Rather, it stemmed from the fact that he had married Rima, socially perceived to be his (or like a) sister.

recounted his story to me. Aside from the fact that he still desired the other woman, he was not sure if Rima wanted to marry him. In order to avoid a commitment, Zuhayr took advantage of his membership of the Communist Party to go to Beirut, where he fought alongside his comrades for about two years during the war. But on his return to Arsal, his uncle pressed him again. This time, Zuhayr felt obliged to accept, given that by now Rima herself was willing to accede to her father's wishes. As Zuhayr explained, like all advocates of cousin marriages, his uncle wanted to keep his daughter within his own house. Based on their shared past and familiarity, coupled with the fact that his uncle was an authority figure to both, the matter seemed straightforward. They were to build on their sibling-like relationship and turn it into a conjugal one.

For Zuhayr, his uncle's decision seemed like a verdict that was unfair (*dhulm*) to both of them. His doubts about sexual desire were confirmed. 'We used to play together and eat together and she would prepare the bath for me, just like the others [siblings]. She was a sister to me.' Zuhayr explained that being raised as siblings inhibited sexual attraction between himself and his wife. There was no *raghbah* between them. Whereas growing up together provided common grounds for their relationship, it also created rifts and distances. A grocery shop-owner on Zuhayr's street once asked me about his marriage. When I explained that he had married his cousin, she retorted, 'If one is to tell the truth, this is *harām* (forbidden). One *bi-khāwi* (becomes sibling) with someone when they're young, how can he marry her? The Sharī'a would forbid this marriage.' Despite my insisting that technically they were not siblings, being brought up together as siblings was enough reason for this woman to contest such a marriage. The ideal of proximity may, then, be elusive and it can defeat its own principle. Perhaps the contempt arises only when marriage is not a matter of *irādah*.

Eventually, Zuhayr accepted his *nasīb* and his marriage to Rima lasted, not without complications, for 27 years until Zuhayr's death in 2005. Although they had eight children together, their relationship was tumultuous as they were very different people with different worldviews and ambitions: he had always been politically and publicly active, whereas she was the opposite. At a certain point, Zuhayr found refuge away from home, in his environmental activism, causing his wife to become ever more doubtful of his attachment to his family while being eaten up by jealousy as they drifted apart. Until his eldest daughter Jumana became engaged to a relative, Zuhayr's narrative, as well as his evaluation of his marriage, had been grim. Without ever forgetting how good-hearted his wife was, he had always thought of his marital relationship as a failure. But later in life, he appreciated his uncle's philosophy in imposing this marriage on him. For after all these years, 'no matter what has happened between my wife and me, we remain, above all, *awlād ʿamm* (cousins),' sharing a past and a history. I have drawn on this history to shed light on the background that prompted Zuhayr and Rima to engage with each of their daughters' desire to marry particular individuals.

3.2   Zuhayr's Daughters

Zuhayr's open-mindedness, and the personal suffering he had experienced in his own marriage, encouraged the hope that his daughters would choose their own marriage partners. In his heart, he wished them to reject the new domestic ideal and had high hopes that they might even leave Arsal. But when Jumana accepted a marriage proposal, engineered by female kin, Zuhayr took his daughter's desires seriously. He knew that she must have already wanted her suitor, Ramiz, before the formalization of the relationship. He was distant kin, and now Zuhayr had to do his homework, to ask around the town about the family's reputation. His principal concern was that Ramiz should build a house for his daughter, the nominal statement of her independence from her in-laws, and support and encourage her to complete her university degree.

Two years into the engagement, Ramiz's behaviour was causing concern. Zuhayr found him childish and 'unqualified' (*ghayr mutaʾahill*) for the journey of marriage, to use the local phrase. Although Ramiz had joined his father's business in Qatar, he never managed to save enough money to finish the house. Instead, he returned to Arsal and wasted his money on buying a fancy motorcycle, riding around the town, and showing off. In the third year of their engagement, Ramiz asked Jumana to leave university and her secure teaching position at the school. To add insult to injury, Ramiz's mother leap-frogged over Zuhayr and asked an elderly relative to 'wrap up' the saga and marry the couple off as soon as possible. Jumana could move in with her in-laws until Ramiz finished the house.

By then Jumana's parents were anxious about her future and Zuhayr asked her to end the relationship. In fact, he even wondered why a woman like her would want to be married to a man like Ramiz. 'I love him!' she responded and stormed out of the room crying. Conscious of appearing old-fashioned and betraying his progressive views, Zuhayr could not bring himself to force her out of love or to get her to grasp the inadequacy of love in securing her future. Never losing sight of her right to consent and to decide on her own marriage, he changed his tactic by assuring her that he would not break their engagement but said he expected her to be less passive about her rights. One evening as we sat around the stove, Zuhayr kindly but firmly rebuked his daughter as she bowed her head and stared at her feet.

> Listen Jumana. These people are compromising your rights. As your father, it is my duty to ensure that all your rights are protected. If you don't allow me to negotiate for you and regain some of your pride, then, believe you me, I am ready to put you on a tray [as one does with coffee] right now and present you to your in-laws. Ramiz and his family don't want to give you a house, they've made false promises. Look what they've done to their other daughter-in-law. They convinced her to move with them until they fixed her house. Ten years passed and she remained a guest. She wasn't even comfortable inviting a friend for tea. After all this time she moved back to her parents' house and told her in-laws she would only return when they finished her house. If this is the way you want to live, I will personally take you to them.

Breached rights, humiliation, disrespect, mistreatment, and 'wasting one's life' were the dangers that Zuhayr, like other Arsalis, believed his daughter needed 'protection' from. Ramiz's family failed the ideology of kinship in this respect, so much so that Ramiz was now *gharīb* (strange, distant) to Zuhayr. But for their part, Zuhayr and his wife did not want to be held responsible for 'standing in the way' of a *nasīb* that their daughter desired. Equally, the way her engagement affected her parents hurt Jumana, who wanted to live up to their expectations. She felt torn between acting as a conscientious daughter and as a loving, understanding fiancée. Eventually, Zuhayr's ideals of respecting his daughter's choice and carving out her own *nasīb* overwhelmed the negotiations. Jumana remained engaged to Ramiz until after her father's death and, as Zuhayr had feared, it took more than five years for her, her husband, and two children to move in to their own house.

Even while the difficult engagement of Jumana was still unresolved, their second daughter, Fatima, was presenting them with a new challenge – one that honed Zuhayr's evaluation of his own marriage and the role of his uncle. Fatima spoke to her parents about a young man she had known for two years. He was from a different lineage, and a further complication was that his grandfather had clashed with Zuhayr's uncle a long time ago when their two lineages were in a state of high enmity. The proposal came when Zuhayr was trying to persuade Fatima to accept his nephew in marriage. But she rejected him out of hand, claiming that 'he had proposed first to my cousin then to Jumana then to me. He wants any girl, not me. I do not want him!' Zuhayr was toying with the idea that perhaps his brother's son would be a better suitor, considering the lack of respect he had seen from Ramiz. After all, his own relationship with his uncle was a primary reason that he had held on to his own marriage despite its shortcomings. His daughters' protection, as he came to understand it, rested on the extent to which he had authority over their partners. Whereas in Jumana's case, his authority was undermined because of a socially distant kin relation, with Fatima he tried to put forward a cousin over whom he was certain to have control. Comparing these cases to Khadija's marriage, described earlier in the chapter, perhaps the same principle stands: that men were looking to forge relationships with men, be it a brother or father-in-law. The difference, though, is that a woman like Khadija had no room for negotiation and hence suffered an unfortunate *nasīb*. But Fatima, like Jumana, established her right to 'live her own life' and to own her *nasīb*. She insisted Zuhayr gave her chosen suitor a chance.

Anxious about having to inform his brother of his son's rejection, Zuhayr explained that 'these days, one cannot *yajbur* (oblige) anyone with marriage. The girl *ikhtārat* [his daughter had made her choice] and there is nothing we can do.' A couple of months later, Zuhayr had no regrets over his decision because he got on well with his new son-in-law and his parents. Moreover, the success of the relationship and the approval of an interlineage marriage that could have been contentious boosted his self-esteem as a progressive man with a consistent ideology.

Through this account, the processes of intergenerational negotiations are illuminated. The way Zuhayr regarded his own marriage sheds light on his and his wife's 'temporal consciousness' as they reflected on their daughters' desires in relation to their own past, and as they accepted the rights of their daughters to make their own choices. Equally, their daughters deployed the vocabulary of modernity – choice, desire, love – in articulating their wishes and in making a case for how these would lead to understanding and a good partnership. These examples point to processes of accommodation and adjustment involving marriage negotiations rather than individualistic drives that challenge parental authority, which can occur when negotiations fail.

4   When Negotiation Fails

The social sanctioning of marriage, primarily by the couple's parents, is a moral necessity in Arsal. There are cases, however, when parental authority is challenged by an extreme act of defiance that tarnishes the reputation of extended families, sometimes for years. Elopement, *khatīfa* (literally kidnapping),[7] often takes place when negotiations, or even prospects of negotiation, fail.

---

7 See Borbieva (2012) for an excellent theoretical discussion of kidnapping and elopement.

This practice is considered the ultimate form of selfishness and attests to the limits of 'individual desire.' Aware of the socially damaging repercussions of elopement, an individual is still seen to choose to follow her desires, especially *raghbah*. The customary form elopement takes begins when a man, with support from his kin or other relations who will act as witnesses to the religious marriage, 'kidnaps' a woman and takes her away from the town (c.f. Marsden 2007). The disgrace stems primarily from the assumption that a woman's chastity is at stake since sexual intercourse is used as a means of pressure to hasten the religious contract in order to save the honour of a woman by making her a legitimate wife. Once a *shaykh* draws the contract, a message is sent out to the woman's parents. Mediators are sought to begin a reconciliation process with the bride's kin, which involves a brideprice or compensation money. Sometimes, parents reconcile with daughters immediately, partly to water down the public shame. At other times, reconciliation never takes place and the couple is outcast.

There are variations in elopement and the extent to which parents are willing to forgive their children. For example, one woman's parents cut her off and forebade all her siblings from talking to her. Because the family was in financial hardship, the daughter had found work baking bread on a stall in a Christian village. There she fell in love with a shepherd and eloped with him. By marrying a Christian man, she had broken a very strong taboo,[8] so much so that no one was permitted to mention her name in the family house. Her isolation lasted more than twenty years until finally the parties reconciled. In another case, a woman eloped with her sister's fiancé a few weeks before her own wedding. She moved to the lowlands with her new husband where they lived for ten years away from their families. Although the parents reconciled with their daughter two years after she eloped, the embarrassment caused was too powerful and the couple felt that it was best to stay away from the town's social life. This kind of elopement, therefore, can be seen as an expression of defiance against parental authority as well as social transcripts that define the sociology of marriageable partners.

Elopement can also be used as a form of politics that bypasses the modern prerequisites for marriage – a lavish wedding, a trousseau, and a finished house – particularly when the person one is eloping with is not disapproved of. At times, parents can be complicit in using elopement as a way to bridge the waiting gap imposed by social expectations. Ali, for instance, had been

---

8   While it is acceptable for a Muslim man to marry a Christian woman, since Islam allows Christian and Jewish wives to practice their monotheistic religions in a marriage, it is unacceptable for a Muslim woman to marry a Muslim man.

engaged to a woman for two years but the house he was building was not yet complete – although he felt that the house was ready for the couple to move in, considering that the essentials were ready: the kitchen, the bathroom, and one living room. But the bride's mother was adamant about not having the wedding that summer and insisted that they postpone it for a year. So Ali 'kidnapped' his fiancée and took her to his uncle's house, where she waited the whole day while the uncle tried to negotiate with her mother. Fearing a scandal, the mother yielded to the pressure and allowed the wedding to take place three days later.

Elopement marriage, then, can be seen as a practice that foregrounds *raghbah* and *irādah* and transgresses intergenerational negotiations around safeguarding the pillars of marriage – partnership and understanding. This kind of transgression of social and cultural norms, however, comes at a price that, in its extreme, threatens to break kinship and social ties, in spite of the triumph of romantic love and desire. The breaking of kin ties, moreover, is a predicament that can cause sadness and isolation. In the next chapter, I explore the ambivalence in kinship, its contradictory emotions and intensities, and the moral registers within which it operates.

CHAPTER 6

Suspicion and Scorpions: The Morality of Kinship

Um Yusif was not herself one afternoon when I dropped by for a visit. Her breathing was heavy and the atmosphere was tense as she watched her son Yusif shuffle through a pile of papers while he pushed the buttons of a blue calculator. When I greeted him and asked how he was, he answered, without looking at me, in an unusually grim tone, 'I feel oppressed by life (*maqhūr min al hayāt*).' I figured this was related to problems he was facing at work. Not really recognizing the severity of the matter, I asked if Um Yusif had one of her usual headaches. 'No, but I am like my son, oppressed by this life,' she replied. By now, I felt inconsiderate for not having noticed how upset she looked. She stood up and frantically started picking up things from the floor: pencils and school notebooks scattered by her grandchildren, orange peel and tissue paper, which she dumped in the stove. As she did this, she continued vociferously with the conversation that my arrival appeared to have interrupted.

> They are the ones who came to my house and begged me to let Yusif become their partner. I didn't go to them. I didn't even want him to work with them. He was doing so well with his pick-up truck ... He's never been humiliated (*tbahdal*; col.) like this ... Now they fire him? ... Shall I just go to the *darak* (gendarme) like other people would? I could complain to the *darak* and reclaim my rights with my own hands.

Yusif and I sat silent, but I was at a loss. Throughout my six years of friendship with the family, I had never once seen Um Yusif as angry as this. While I couldn't grasp exactly what was going on, I deduced that she was referring to her brothers. This attack on her natal family, and the mention of the gendarme, clearly meant that there was a serious problem, a bitter disagreement that had taken place between Yusif and his uncles at the stone-cutting factory. I asked, 'What exactly is the *khilāf* (disagreement, conflict) about?' But the response, synchronized by Yusif, his mother, and his sister Hana, who was stepping into the living room with a pot of tea, was as incomprehensible as the story that subsequently emerged in spurts: 'Conflict? No, no! No conflict (*khilāf? La la, ma fī khilāf!*)!'

Um Yusif had a soft spot for her eldest son, Yusif. His *nasīb* (fate), she thought, never matched his kind personality and potential. His future seemed bright when the Communist Party sponsored his studies in the USSR in the

1980s. But funds became limited before he could finish his degree. Upon his return to Lebanon, Yusif enrolled in a Fine Arts programme at the Lebanese University. But there, too, he failed to achieve his goal. Yusif fell in love with a classmate from the valley halfway through his programme and eloped with her against her parents' wishes. Soon after, the couple had children and found themselves caught up in the everyday business of making ends meet.

Um Yusif invited Yusif and his new family to move out of their flat in a nearby village and into the family house while they all cooperated in building a second floor for Yusif's family. Um Yusif gave him some money and borrowed more from friends and relatives to buy a second-hand truck so he could set up a business transporting different goods (rocks, mainly, but also sand, wood, cherries, and other agricultural produce). Yusif worked hard and often had to sleep in his truck for several nights if his clients were outside the Biqaʿ area. From the look of his truck, you could easily tell that most of his money was spent on mechanical maintenance. His life choices had forced him to settle for harsh work when he could have been a famous painter; at least this is what his mother thought. Nonetheless, he got by with the help of his mother, sister, and wife, who had a secretarial job in the town.

When Yusif's uncle Fahd was in the process of establishing a stone-cutting factory with his brothers, he approached Yusif to become a partner with a 10 per cent share. This arrangement meant that Yusif would not get a daily paid wage like the rest of the labourers. Rather, he would reclaim his share only when the factory started to make a profit. Yusif's mother and father, who lived in the south at the time, were sceptical, bearing in mind their son's commitments, as it was not clear at what point he would start earning money. But Fahd persuaded them by drawing on the common ideology that postulated *al qarīb awla min al gharīb* (kin have priority over strangers). On the one hand, this presupposes that kin, by definition, can be trusted and will not, as the locals say, 'eat [cheat] one another.' On the other, it presumes that if anyone was to benefit, then priority should be given to kin.

In the light of this principle, Yusif accepted the offer and worked for two years, mainly operating the machines in the factory but also doing a number of other physical tasks. In those two years, the factory made hardly any profit, or so Yusif was led to believe. He gave the factory use of his pick-up truck, as well as his time and effort, without ever complaining. One day, his mother's brother Fahd brought in his own son and asked him to start operating the machines. Was Yusif about to be dismissed? The move seemed to indicate that he was. So he went home and he waited there despondently, hoping that it was all a misunderstanding. But two days passed without Fahd or any of his other uncles contacting Yusif. His suspicions were confirmed: they had discharged him without even telling him so to his face, and without giving him any of the money they owed him. All this happened in December, a very difficult month to find work in Arsal. Yusif and his mother were left nursing a lingering feeling of betrayal.

Yusif very quickly managed to find an alternative source of income. He bought a used van and a licensed car plate number allowing him to use his vehicle for public transport and started to drive university students to the city of Zahle. Moreover, soon after getting back on his feet, he visited his uncle Fahd. 'Blood does not turn into water' ('Blood is thicker than water'), he concluded after he returned. Um Yusif, however, was less forgiving as this incident conjured up deeper pain related to other family injustices. Unlike her son, she could not allow her wounds to heal and move on. In the meantime, Um Yusif felt she was being pulled in opposite directions. Her ethical sensibilities, coupled with pressure from her children, demanded she reconciled with her natal family. But at the same time, she had a strong urge to make a statement by distancing herself from her kin. Um Yusif was trapped in an ambivalent situation, 'the simultaneous coexistence ... of ... contradictory emotions or attitudes towards a person or thing' (Peletz 2001: 414).[1]

This chapter focuses on moral ambiguity in kinship relations. We have seen in previous chapters how the family is considered a haven for sameness and solidarity and a tool that 'contains' its members. We also saw how it is simultaneously a site for conflict of interest and difference. Discourses of the family as a haven for love, harmony, and sameness tend to promote a utopian vision of intimacy. But everyday practices often counter this image, leading to a dystopian vision of 'anti-kinship' in which 'the positive attributes are entirely replaced by selfishness [and] loneliness' (Lambek 2013: 243). In this contradictory appraisal, kinship becomes at once a site of refuge and suspicion. I draw on Um Yusif's story to elaborate the acts that rendered her relatedness to her natal family 'anti-kinship.' The sections that follow consider the idioms of suspicion that acknowledge breach and violation as constitutive aspects of kinship. While betrayal, mistrust, and disappointment are intrinsic to kinship, as they are to human nature, my interlocutors situated such attributes in modernity and economic change that seemed to render humans greedy and selfish. This was seen as a challenge to human relations operating within registers of morality that demanded kinship continuity and harmony. Through the labours exerted to recalibrate kin relationships and emotions, Um Yusif's story articulates

---

1 Peletz resorts to the Oxford English Dictionary to distinguish analytically between 'ambivalence,' 'ambiguity,' and 'diffidence.' For him, ambiguity refers to a cognitive rather than emotional phenomenon, whereas diffidence denotes a shyness or a 'reluctance to express one's emotions, attitudes, or self' – which could, he argues, be mistaken with absence of affect (2001: 415).

what people thought kinship ought to be and why her family might have denied or played down the possibility of conflict. The chapter therefore captures Um Yusif's 'moral breakdown,' that moment when a person 'step[s] away from their unreflective everydayness and think[s] through, figure[s] out, work[s] on themselves and respond[s] to certain ethical dilemmas, troubles or problems' (Zigon 2007: 140). By exploring how women from different positionalities and trajectories envisioned the risks of proximity, the chapter highlights the intertwinings of gender and kinship over time (Collier and Yanagisako 1987). Attention to gender and generation renders visible the power imbalances that are entangled with the dynamics of social change in Arsal.

1 Ensnaring Brothers and Suspicious Sisters

The term *ahl* (kin) predominantly encompasses positive qualities. It presupposes trust and takes for granted reliability among kin. Other relationships (friendship, neighbourliness, colleagueship etc.) aspire to replicate kinship, as kin relations are considered far superior in terms of their quality, longevity, and natural strength. Yet, there is an inherent contradiction in this ideology that happens to recognize the potential failure of kinship. It is precisely the high expectations of kin that pose the challenge of fulfilling them. The shock expressed at a failing father, for example, is counteracted by an obvious familiarity with such failure. 'Your biggest enemy is your father and your brother (*akbar 'aduwwak khayyak wa bayyak*),' Um Yusif would say when she spoke about the problem with her brothers. But why are fathers and brothers, who are supposed to be the most reliable, depicted in such a negative light? 'The proverb makes no mistake,' Um Yusif asserted once. 'If you meet someone on the road and then they hurt you, you forget and forgive them an hour later, but not the close ones.' The suggestion here is that there is danger in proximity; the local ideology of kinship carries room for breach. This, however, does not prevent the pain that ensues from such failure. Other anthropologists have come to the same conclusion:

> 'Family' are those who owe one loyalty. Ideologically this is not open to doubt. Nor is it subject to empirical falsification. Betrayal by a brother is so much the worse because a brother betrays. Assistance from an uncle is so much the more welcome because an uncle assists. The statistical probabilities of either do not affect the response; nor do they affect the morality on which that response is based.
> JUST 1991: 130

Idioms of suspicion in relation to kin are common and circulate in different versions in Lebanon. For example, some express suspicion towards kin through the rhyming adage *al-aqārib ʿaqārib* which translates as 'kin are scorpions.' Like scorpions, kin can be treacherous. They sting when one least expects them to, leaving the victim with a lethal injury, albeit an emotional one. Although the Arsali expressions seemed to single out male relations, the father and the brother, such idioms were used in common parlance in a gender-neutral manner and were invoked as metaphors of wariness as far as kinship and close relations in general were concerned. Yet, with the women that I worked with, there was a specific relationship that carried more tension than others. While *al-akh fakh*, 'the brother is a trap,' captured the rifts that could inevitably come about between brothers, it expressed a specific apprehension that sisters felt towards their brothers in post-war Arsal.

The significance of the brother/sister dyad has been noted in the Lebanese and broader Arab contexts (Joseph 1994; Jean-Klein 2000; Baxter 2007). Suad Joseph (1994), for instance, identifies cross-sibling relations as a site where self-formation and socialization take place through the 'love-power dynamic,' a move away from the parent-child dyad as a locus for such processes. 'One's sense of self,' she argues, 'is intimately linked with the self of another such that the security, identity, integrity, dignity, and self-worth of one is tied to the actions of the other' (1994: 55). A boy's sense of self, Joseph contends, is realized by loving and controlling his sister. By contrast, a girl's sense of self is achieved through loving and submitting to her brother. Recursively, sisters and brothers cooperate on a daily level to create 'relational selves.' It is this kind of connectivity, Joseph argues, that reproduces certain forms of Arab patriarchy – 'the dominance of males over females and elders over juniors (males and females) and the mobilisation of kinship structures, morality and idioms to institutionalise and legitimate these forms of power' (1994: 55).

I concur with Joseph's argument that subjects related by kinship are 'enmeshed' with other subjects. Like the Beirutis she describes, my Arsali interlocutors saw the brother/sister relationship as one of the closest and most valued. My ethnographic material, however, interrogates the power dynamic as depicted in Joseph's Camp Trad, where there was an acceptance, almost an invitation, to reproduce patriarchy. Flaur, a character in the Camp, actually 'enjoyed' submitting to and being disciplined by her brother (1994: 51–52). As Joseph notes, 'connective' relationships can also be hostile. In the Arsali case, while the brother/sister relationship was invaluable, and while idioms of 'compassion' and 'protection' were used to characterize this relationship, I found among my female interlocutors, especially the unmarried ones, a growing scepticism surrounding discourses of brotherly protection. To an extent, many unmarried women I spoke with sought paid labour precisely to avoid female dependence on brothers. To unpack the gendered aspect of *al-akh fakh* that occupied the imagination of unmarried women, I take Joseph's (1994) invitation to explore the brother/sister dyad and family dynamics more generally in different cultural and historical contexts. In the following sections, I locate expressions of suspicion towards the brother in hegemonic models of femininity and personhood that were articulated in the rural post-war context of Arsal, when economies, households, and kinship structures were undergoing transformation.

1.1   Female Altruism

My interlocutors explained to me that *tadhiya* (sacrifice) constituted a key element of kin loyalty. But *tadhiya* was gendered. While a father was expected to sacrifice (*yu-dahhī'*) by giving up his time, labour, and health for the wellbeing of his children and family, *tadhiya* constituted a key aspect of female selfhood. At different stages of their lives, women were expected to make sacrifices for their families, but their sacrifices seemed more extreme and ranged from giving up social to legal rights. For example, an unmarried woman might be expected to delay her marriage or not get married at all in order to look after an elderly parent. A son's education might be prioritized over a daughter's. Women were also expected to relinquish their rights to family inheritance.

These discrepancies emerged in the conversations I had in the carpet cooperative when we discussed women's rights and their awareness (*wa'y*) of these rights. The women questioned the boundaries between exercising female *tadhiya* and naively giving up rightful privileges. My interlocutors felt that Arsali society and its men were sometimes unfair, but observed that more often than not, women were complicit in reproducing their inequality. They often neglected their rights, especially when these rights were legally and religiously sanctioned. For example, according to Islamic Sharī'a, women are entitled to inherit half the share of men. Arsalis were aware of this fact. Yet, women almost never actively asked for their share of inheritance. Nor did their families necessarily volunteer it to them. When I asked a herder about inheritance practices in his family, he sheepishly confessed that, 'The truth is we are not very fair because we *tannish* (col. ignore) daughters' rights. It is *harām* because Islam says we should give [their share] to them, but we don't. Among us, a generous man might give a daughter two or three heads [of sheep], but it is unusual.' The situation was more complicated when it came to land inheritance.

In one of our discussions during a tea break at the carpet workshop, Khawla told us about her father's sister, an unmarried woman in her fifties who unexpectedly demanded her share of her inheritance from her father. Surprised by this unconventional request, her pious brothers conceded their sister's rightful demand. But just before their trip to the official clerk who was to formalize the land transfer, their sister changed her mind, thus reverting to the expected mode of female altruism. Why would her aunt demand her rights in the first place if she were going to relinquish them, I asked. Recognizing the contradiction, Khawla felt that her aunt may have been 'testing the limits.' 'In our context,' she explained, 'it is abnormal for a woman to demand her rights. Even *I* would give land back if my brothers gave it to me!' Khawla's reference to herself was offered as proof of the extent of women's *tadhiya*, for she was well aware of her reputation as a strong and outspoken woman who did not suffer fools, both within her family and among her co-workers.

The relinquishment of women's rights in spite of the Sharī'a law is not an uncommon act, though it may not always be entirely disinterested. Other anthropologists have shown how, by renouncing their right to land, women make claims over males and even strengthen their positions through morally indebting their male kin (Peters 1980; Abu-Rabia 1994; Granqvist 1935). This kind of trade-off calls to mind Denise Kandiyoti's 'patriarchal bargain' (1988; 2005) by which women sacrifice certain rights and short-term gains in return for long-term security. In the case of unmarried women, the compensation entailed the continued guardianship of the brother after the demise of the parents. This arrangement was all the more pertinent in a context where it was not customary for women to live by themselves and where the job market was predominantly masculine and limited. Perhaps Khawla's aunt wanted reassurance that the bargain was on-going, that the exchange was still valid, and that when the chips were down, her inheritance rights would be acknowledged.

Parents generally relied on the bond with brothers to safeguard the social, emotional, and even financial needs of their daughters after they were gone. The only property that was commonly given to single women who were beyond marriageable age reinforced the 'bargain,' as it represented a guarantee of the bare material minimum: a roof over their heads. I was told that fathers looking to ensure their unmarried daughters' rights would sell them their family house for insignificant amounts (such as the equivalent of one USD). Such arrangements sidestepped the Personal Status Law[2] of inheritance – which in this case

---

2 A law based on religious laws, Personal Status Law differs from one religious sect to another. Out of Lebanon's 18 religious sects, 16 (apart from the 'Alewites and Isma'ilis) have their own personal codes. Matters considered 'personal status' include marriage, divorce, custody of children, and inheritance. Each religious sect has the right 'not only to administer its own affairs, but also to legislate, judge, and carry out sentences in matters pertaining to their respective congregations' (Shehadeh 2010: 212). This has been a continual issue of concern for feminist activists in Lebanon.

would follow the Sunni Sharīʿa – that could disadvantage women by granting them half a share, or could complicate inheritance processes that required divvying up a house. They also helped avoid a situation where women felt under social pressure to give up their rights, in principle at least. The women I spoke with explained that even when such measures were taken to safeguard women's rights, sisters still tended to share parental houses with a struggling younger or older brother looking to marry and make a home. Women without brothers shared parental homes with other unmarried sisters, or even lived alone – a state that was at the heart of constructions of female vulnerability.

1.2   God Did Not Say 'Live Alone'

An appreciation of the central and diverse roles that women played in their families challenges the ideal of female vulnerability. From household management to running their families' small businesses, to undertaking agricultural and herding activities, to dealing with hospital bureaucracies and even utility officials, women showed anything but vulnerability in terms of facing everyday life. Unmarried women were known (or even expected) to take on the responsibility for their families. Um Yusif's own daughter Hana is a case in point. As a craftswoman, she developed artisanal expertise in a variety of areas: carpet weaving, textiles, jewellery, and embroidery, among others. Hana became an instructor at one of the local NGOs and in more recent years established a small weavers' cooperative. She embarked on collaborative projects with international NGOs, ministries, and even a prisoners' rehabilitation programme in Beirut. Her work took her to Arab and European capitals where she participated in international exhibitions; all this without any language skills and with only elementary schooling. Hana used her income to help build Yusif's house, refurbish her own parental home, and frequently lent friends money. While not every woman was as skilled as Hana, the tasks and responsibilities undertaken by unmarried sisters, especially as they entered 'spinsterhood,' were remarkable.

Ideas of female vulnerability, then, may relate to more than just the ability to handle life. I suggest that they emerged from a cultural fear of being alone. This 'unthinkable' (Jamieson and Simpson 2013; c.f. Allerton 2007) notion illuminates the importance of relationality in Arsali sociality. Being alone disavows the intimacies and proximities that are at the heart of togetherness (Chapter One) and is a state to be pitied, both in the long and short term. People showed concern about someone who expressed a desire to be alone, even for the mere want of temporary space. Something *must* be wrong with a sister who left the family gathering only to be found sitting by herself in another room. My friends constantly tried to talk me out of my incomprehensible living arrangements, in a room on my own in the spacious local NGO. 'God did not say, "Live alone",' they would insist, implying that being alone and unsociable defied human nature as God intended it. When I slept at friends' houses, it was always with other people in the room (with four sisters on shared mattresses, in one case). Eating alone was just as sad, and people who travelled and found themselves having to snack or eat without company described a loss of appetite and a lack of joy in food.

If being temporarily alone was a cause for concern, being permanently alone – the presumed state if one remained single – was a cause for great sadness. Both men and women 'fulfilled their fates' through marriage and the creation of a family. Failure to realize this goal led to a sense of flawed subjectivity for both genders, a feeling of not fulfilling one's purpose in life that abandoned people, literally, 'to their loneliness' (*li-wahduhum*). This was Khulud's predicament. She was the single daughter of an elderly woman and they lived together in the family house until the mother died. People felt sorry for Khulud, who was in her mid-thirties and past marriageable age by local standards, and worried about her 'being alone' (*li-wahdaha*). Socially, one could hardly say that Khulud was left alone. In the first few weeks, neighbours would send one of their children to sleep next to her so that she was not alone at night. From time to time a cousin from another neighbourhood would spend a couple of days at her home to keep her company. People regularly checked on her and relatives insisted she ate with them and stayed the night when she dropped by for a visit. Her house was in a courtyard shared by four other houses, and her neighbours were fully involved in each other's lives. Yet Khulud was sad. As she told me, 'With us here [in Arsal], a person has no worth without a family, especially women. This is my fate.' Eventually, she got a job as a carer in a Sunni orphanage in Beirut. That year it was accepting applications from several unmarried Arsali women, and one of her cousins encouraged her to apply, on the grounds that she would find kin and friends in the organization while earning an income. Her job covered food and accommodation in a dormitory, and Khulud could visit Arsal once a month to check on her house and see friends and kin.

Drawing on religious vocabulary, Arsalis promoted an ontology of social relatedness in which individuals must enmesh themselves with others. Whether by will or by fate, a future of aloneness was seen to reduce life to 'just living,' rendered as *'ishah min qillat al mawt* ('life made possible only by the lack of death') – in other words, a life impoverished by the lack of relationality. There were social mechanisms in place to ensure that people did not end up alone. Moral obligations dictated that children looked after their parents in their old age, especially if a parent was left alone through the death of a spouse. This duty usually fell on the shoulders of an unmarried daughter, or on the eldest son, if the other children were married. For a son, the chances of (re)marriage remained open till late in life. For daughters, spinsterhood came early and single women feared their imminent transformation into a burden (*'ib'*). Children were expected to outlive their parents and images of a daughter's aloneness caused much parental concern. The reliance on the goodwill of the brother, in such a case, counteracted these worries. Arsali single women, however, were wary of the significance and implications of men's goodwill.

1.3   Cinderella – But without the Prince

A key distinction was made between the role of the father and the brother vis-à-vis an unmarried woman. Whereas fathers were expected to look after their daughters as part of their 'natural' roles, protection by brothers was perceived as *always* at risk. As much as a brother loved his sister, there would come a time, women believed, when his loyalties would be tested as they became divided between his sister and his wife; the latter evidently assuming a higher status. In the contemporary aspirations of home-making, in which a woman 'entered her *mamlaka* (kingdom)' (Chapter Four), there was no room for more than one reign. This was, of course, very different from the way Arsalis lived only a generation ago. When a new wife moved into an extended household, she felt vulnerable and needed to prove herself to the other women (mother- and sisters-in-law) before establishing her own authority. Having another woman in the modern *mamlaka* could cause tensions and threaten the autonomy a wife was expected to enjoy. Even if the sister-in-law was employed or had her own income, a wife was believed likely to covet all the attention and affection her husband gave her.

While this scenario of female competition was plausible, it was pushed further by an exaggerated fear of the potential actions a wife might take to assert her dominance and turn the situation to her advantage. While she would be expected to respect her husband's obligations toward his kin, my interlocutors contended that she would also be looking for ways to make her own life easier. How better than by exploiting the sister-in-law? This fear of exploitation was expressed in terms of the excessive household labour a wife might impose to free herself of these duties so as to live a life of comfort. Women used '*isti'bād*' (enslavement) to articulate the degradation an unmarried woman could expect as a result of her dependency on her brother. This subject was a particularly sensitive one in the carpet workshop as all the women who worked there were unmarried and most were 'spinsters' by local standards. It particularly rang true with Nihal, whose relationship with her eldest brother Anwar was known to have broken down over the years. His behaviour had brought grief and humiliation to Nihal's family.

Anwar, his wife, and their four children lived in the same house as Anwar and Nihal's parents, though they had their own private wing, linked to the rest of the house by a single corridor. However, his parents did not have a harmonious relationship with their daughter-in-law, which made things difficult for them. From the beginning, they felt that Anwar's wife was too demanding and materialistic (*māddiyyah*) and that she put pressure on Anwar to make quick money. They blamed her for encouraging her husband's involvement in illegal activities brokered by her male kin. When Anwar was caught by the Syrian authorities and accused of smuggling drugs across the border, his family found themselves having to look after his wife and four children on top of handling his complicated legal case. This dragged on for over fifteen years, given the severity of the charges against him. Nihal's elderly parents, who had been relatively well off before their son's arrest, began spending all their savings on his case. Over the years, they were forced to sell off their land and soon found themselves borrowing huge amounts of money from kin and friends to cover Anwar's legal fees, the cost of their frequent trips to Damascus, and the bribes that had to be paid to navigate Syrian bureaucracy. They became desperate and, eventually, both fell sick with stress-related ailments – the father with diabetes and the mother with severe eczema. 'Anwar broke us (*kasarna*)!' his mother would complain to me – an expression that suggests both physical and financial ruin.

As Nihal and her siblings matured, they found themselves inheriting the responsibility of Anwar's case and family. Three of her older brothers were married with children, and though they still contributed towards Anwar's family, the financial burden fell mainly on Nihal, her four unmarried sisters, and one unmarried brother. The brother worked as a labourer in one of the quarries and the sisters took different jobs based on their skill sets. Of the five daughters, two worked in the Beiruti orphanage mentioned earlier, Nihal in the carpet workshop, and one stayed at home to look after their poorly parents. It was Nihal's eldest sister who felt most aggrieved for, unlike her sisters, she had left school at a very early age and had only basic literacy. To help her family out, she found work in one of the mountain resorts, about an hour and half's drive away, where she baked fresh bread in a restaurant. Her sisters and mother told me how much this new reality saddened her. As the eldest daughter, she was like a mother to her siblings and they grieved that she was compelled to bake bread for other people and live away from home in another town, instead of being able to care for her own family.

What irked Nihal particularly was that Anwar's wife expected her sisters-in-law to work to provide for her and her children. She refused to work herself and, according to Nihal, would often feign illness, taking advantage of the fact that Anwar's family would never abandon their moral commitment to his children. The burden of Anwar's case took over the lives of Nihal and her siblings. Her younger brother could see no way of building a house for a future wife when all his earnings were spent on Anwar's children. He feared that, in contemporary Arsal, no wife would agree to live with his parents. The sisters felt that they could not abandon their younger brother and parents to cope on their own with the financial burdens Anwar had imposed on them, in spite of the fact that Nihal's parents wanted their daughters to marry and 'fulfil their fates.' They were enacting the kind of female altruism discussed above, putting the needs of a brother (in this case Anwar's family and perhaps the younger brother too) above their own. Although time began to heal some of the tensions between Anwar's wife and his family, when Anwar was finally released, his behaviour towards his sisters gave credence to the expression *al-akh fakh* (the brother is a trap). Nihal complained that Anwar's wife felt empowered by his release and began to 'turn him and the children against his own family.' In tears, Nihal told me how one of the children, now in her mid-teens, shouted insults at her grandfather and even told him he should live elsewhere. On more than one occasion, arguments escalated and Anwar became violent towards his sisters. The parents begged Anwar to find a house and move away so that their relationship would not break down entirely. While Nihal and her family accepted that Anwar's wife was not the kind of woman they had wished to join their family, their sense of betrayal came from Anwar himself, who showed no gratitude for the sacrifices they had all made for him over many years.

Although these details were not openly mentioned in the carpet workshop, Nihal's colleagues and others more widely in the town were aware of the outlines of Anwar's story. Without having to say anything, Nihal's co-workers knew that she embodied the pain caused by a brother's treachery. Out of respect for Nihal, and to lighten the mood of the discussion we were having in the workshop, one woman stated that she would rather take a married man, even as a third wife, than succumb to the domineering sister-in-law! Her colleagues teased her by reminding her that she couldn't even secure the status of first wife, let alone third. They decided that her only option was a dying old man who needed someone to nurse him. 'At least I would be in charge,' she retorted. 'Isn't *that* even better than being like Cinderella? At least Cinderella got the prince. We will only get the suffering.' This remark aptly evokes the 'Cinderella complex' in that it describes 'a poor third relation in a triad' (Dent 2001: 25): like Cinderella, the unmarried sister in these narratives is wronged by her sister-in-law. In our conversation, however, the sense of treachery was directed towards the brother – *al-akh fakh*: it is the *brother* who is seen to exercise deception or betrayal, as in Anwar's case. The wife prioritizes her own interests, an ambition that women understood. The brother, on the other hand, is a *promise* of protection, pledged in childhood, only to falter and disappoint at the critical moment when, by selfishly choosing to succumb to his wife's wishes, he takes her side against his own sister.

Admittedly, even to my Arsali interlocutors, this folk theory might seem unfair to all those brothers who looked after more than one household and honoured their sisters, both married and single. It also seemed unjust to the sisters-in-law who developed amicable and affectionate relationships with their husbands' sisters. Most of the women who elaborated on the perfidious brother/sister dyad expressed intense love (*hubb*) towards their brother at the same time. This suggests that the betrayal of a brother is not necessarily an inevitable occurrence. Rather, I see *al-akh fakh* as one articulation of the tension between trust and betrayal inherent in the ideology and practice of kinship. The 'enslavement complex,' as a gendered discourse of brotherly betrayal towards a sister, needs to be understood in the context of evolving gender roles and hegemonic models of femininity that cause inevitable tensions between female altruism and the promises of kinship, where individuals who might have previously sought refuge in kin relations find themselves questioning its presumed nature.

2 Of Failed Bargains

In the above sections, we saw how unmarried women, from the vulnerable position of potentially ending up alone without kin or kin support, questioned kinship's version of itself, or rather their own expectations of kinship. Continuing with Um Yusif's story, this section explores how the tensions inherent in the brother/sister relationship persist after marriage. Um Yusif's use of a saying that I only ever heard her, and no one else, express, *Akbar ʿaduw lil khāl huwa ibn al ukht* ('The biggest enemy for an uncle [mother's brother] is his nephew [sister's son]'), suggests that the potential enmity between a sister and her brother may persist after the woman marries and may even extend to her children, especially sons. The incident with Um Yusif's brother drove her to look back on her life and engage in what David Berliner calls '"reflexivity-in-action" … moments during which self-knowledge and feeling about the self are being produced' (2016: 5). Um Yusif seemed to be re-evaluating her positionality as a woman in her kinship roles of daughter, sister, and wife. For several weeks after Yusif was dismissed from his job, Um Yusif went backwards and forwards over different moments in her life, recalling incidents that charted her 'trajectory' (Ghannam 2013) as a woman and teased out the injustices of social expectations. In this process, she interrogated and critiqued patriarchal practices (and 'bargains') and power dynamics that normalized male privilege, often at the expense of women, young and old. Shedding light on this perspective of an older woman is important, for it is not 'younger women alone [who are] resisting, rebelling, complaining, offering alternative visions of family and social life' (Lamb 2000: 8).

The actions of Um Yusif's brother had multiple repercussions. At a basic level, her son felt hurt and exploited, as he received nothing in return for his contribution. Using the local vocabulary, she complained that her brothers had 'made a mule'[3] out of her son. But these actions affected Um Yusif personally as well, beyond her role as an empathetic mother. The manner in which she expressed pain conjures up notions of connectivity, discussed earlier, in which the dignity, integrity, and self-worth of one self (her son) is tied to another (her self). This was not the first time that one of her brothers had failed her, perhaps a reason why her pain lasted longer than her son's. There had been precedents of disappointment. Um Yusif recalled how once one of her brothers failed to stand by her daughter Hana at a time of need. Hana was visiting her father in the south when he was hit by a car, taken to the emergency unit, and hospitalized for a number of days. As Hana was only supposed to be staying for two nights, she had not got enough money with her to cover the hospital fees. Her well-off uncle visited them in the hospital but ignored Hana's subtle request for help when she made it known that she was short of cash. 'He left Hana to deal with this issue, a girl on her own, without money, far away from home!' Um Yusif cried. Her brother's actions were selfish, she felt. He prioritized his own convenience at the expense of a niece at a time of need. They also reflected a kind of greed that manifested itself in a refusal to share his wealth.

In spite of her feelings of disappointment, however, Um Yusif never confronted her brother over this incident. While unpleasant, it had taken place privately, in the confines of the hospital ward. But the fall-out with Yusif was soon public knowledge. 'We have become the talk of the town,' Um Yusif fretted, as she endured neighbours' questions and comments, some out of concern, others more derogatory: *'Shū? Halakū lu?'* ('Have your brothers shaved him?'; in other words, 'Did they get rid of him?'). Either way, she felt ashamed and upset that her kin had helped tarnish her reputation for being the most conciliatory person in her neighbourhood. She told me that during all the adversities Arsal had endured, she had never once been *za'lāna* (in disagreement) with any of her neighbours, apart from one single exception. During the 1975–1990 war, she

---

3 *'Yu'attil 'ala dhahrih'* – literally 'put a weight on his back,' as is usually done with a mule, or 'treated him like an ass.'

had confronted a woman whose son, a member of a rival political party, had reported Yusif to the Syrian Secret Police. She had been driven to threaten her neighbour in order to save her son but since then, despite her strong dislike for this woman, she had not fallen out with her again. The last thing she expected was that her own brother would mar her image and paint her as someone who quarrelled with her own kin, and worse, as someone whose kin failed to look after. After all, kinship is a 'set of commitments, played out in practice and publicly articulated' (Lambek 2011: 3). Um Yusif blamed her brother, not just for relinquishing his brotherly duties of looking after his sister, but also for breaching his obligation to protect her honour. This behaviour, she felt, was a sign of the time, when sacrificing a kin relation seemed feasible in return for economic gain. 'Humans have always been greedy,' Um Yusif told me. 'From the day God created us. Look at Qabil and Habil (Cain and Abel). But today, it's too much! Life has changed. A brother would sell his own flesh and blood for greed (*tama'*).'

As the days passed and none of her brothers pursued the matter, explained themselves, or checked up on her, she read the situation not just as a failure of kinship (as the hospital incident might have suggested) but rather as a *denial* of kinship. Um Yusif construed the problem between her brother, the primary antagonist, and her son as essentially a conflict between herself and her natal family. As far as she was concerned, her kin had, in effect, 'disowned' her. To protest her belief that 'they refused to acknowledge her' (*lam ya'tarifu fīnī*), she made the decision to 'dis-acknowledge' them, so she stopped visiting her mother's and brothers' houses. Her principled decision, however, soon crumbled against the moral Goliath of kinship. As she endured moral pressure and sadness in the following weeks, Um Yusif began to review her relationship with her entire family and to question retrospectively the gender imbalances that had occurred in her life.

2.1 *It is All the Mother's Fault*

Whenever Um Yusif talked about the family disagreement, now referred to as 'the Subject' (*al-mawdū'*), she often used the plural 'them/they,' thus seeming to implicate all her natal kin (*ahl*) in the category of wrongdoer. Her surviving natal family, however, comprised several people: her mother, four brothers and four sisters, excluding their children. As she sifted through her recollections of family experiences, it became clear that some – her sisters and one brother who had been sickly throughout his life – were exempt from blame. These siblings occupied a similar subordinate position to hers in the household. In a sense, her stories were a commentary on patriarchal practices that favoured sons – and more abled ones – over daughters. Um Yusif took a stand against her mother, who she believed endorsed discriminatory practices, especially against the subordinate female. 'My mother,' she proclaimed, 'is the cause of everything.'

Um Yusif recounted stories and decisions that took place before her marriage, more than 45 years ago, but about which she now felt bitter. She remembered how, from the age of seven, she was made to carry coal, tend to the herd, and take responsibility for household chores. At the same time her brother, who was almost the same age as her, was privileged and given an education that in due course would turn him into a renowned medical doctor. She and her sisters, by contrast, were deprived of any schooling. This kind of gender bias, especially in education, was not specific to Um Yusif but was practised on her entire female generation, to such an extent that one would be hard pressed to find a single literate woman in her age group. But the virtues of education were ever so clear *in the present*. Um Yusif was under no illusion that school education would have gained her formal employment – no woman of her generation had that. But she was aware of the loss of knowledge of the world that came with illiteracy. Um Yusif craved the tools of self-improvement that she could have possessed had she been able to read and write. She would have at least been able to understand her faith better. 'Why do I have to wait for television to learn a verse or two from the Qur'ān when I could have learnt all of it?' The proliferation of satellite television in Arsal since the late 1990s corroborated these feelings of a potential alternative personhood, as Um Yusif discovered her window on the world.[4] She found a new passion for world news and political analysis, documentaries and soap operas that exposed her to different ethnicities, classes, and cultures, and TV stations like the Saudi-owned religious channel 'Iqra' that focused solely on religious education.

While women's exclusion from education could be passed off as a particular gender bias that was relevant to the now out-dated structural inequalities of 'old times' – and one that applied to a whole generation rather single gendered individuals – Um Yusif brought to the surface other prejudices that reflected deeper and more personal injustices. At the heart of her grievance was the way her mother had distributed family resources and, according to her distraught daughter, her emotions too. Years after Um Yusif's father died, her mother decided to divide the family house into shares,[5] but these were unequal. The

---

[4] This resonates with the ethnography of other Arab rural settings. Compare, for example, Abu-Lughod's interlocutors who expressed similar sentiments about television: 'People like me are angry with our parents for not giving us an education. Now, after schools and television, people know a lot more' (Abu-Lughod 2008: 62).

[5] It is unclear whether Um Yusif's mother was the heir, or whether she just accepted the older brothers' decision on the division of their late father's property. Whatever the case, Um Yusif sees her mother as the person in charge.

women were automatically ruled out. Addressing me, Um Yusif rhetorically asked her imaginary mother, 'How can you divide the house among the three of them (the three eldest brothers) when there are nine of us? And how can the doctor inherit the same share as Jamal, who can barely button up his trousers?' Jamal, the youngest brother, was unmarried. He was educated and had a secure job as a civil service employee, but he received a lot of sympathy because he was disabled and used a wheelchair. Um Yusif believed her mother was unfair because she denied her daughters and a vulnerable son their legal and religious claims to inheritance and their moral and emotional rights to a level of comfort a mother ought to provide, to the benefit of her other sons.

Um Yusif revisited other instances when her mother failed to act in ways that expressed appreciation of her. She told me once, when at her mother's house, some guests were discussing trousseaus (*jihāz*). Each woman listed what her mother had given her before her wedding. But Um Yusif had nothing to contribute to the conversation and found herself remarking sarcastically, 'I still have what my mother gave me. In fact her gifts are still covered because I wanted to protect them from dust.' This embarrassed her mother who knew that she had not given her anything. At the time of my research there was no standard rule or custom in relation to the *jihāz*.[6] Parents tended to offer their daughters different items, the most common of which were electrical appliances or a hand-made Arsali carpet, one of the most essential requirements for a house. Women of the older generation, by contrast, received barely anything from their parents. Perhaps what Um Yusif was bitter about was her belief that her mother *could* have afforded to give her something but chose not to, and had watched her struggle to make a home.

> I started my marriage with an empty, unfinished house. We slept on the bare floor for an entire year until I went to a second-hand shop that was selling a ton of second-hand woollen hats for 10 liras. I bought all of them, de-threaded them one by one, and on my own wove this carpet on which we are still sitting now.

Not unlike the unmarried women discussed in this chapter, Um Yusif was critical of the kinship structures and ideologies that favoured sons over daughters and reproduced inequalities. Her age and position in her life trajectory, as a mother with her own adult sons and daughters, granted Um Yusif a fresh perspective on these matters. How she treated her children in the present was in

---

6 For a comparison with other Arab settings, see Abu-Lughod (1986) for a discussion of bride price and see Mundy (1995), Maher (1974), Peters (1980) and Abu Rabia (1994) for a discussion of dowry and bridewealth (*mahr*).

dialogue with how she was treated in her own family in the past. Um Yusif had maintained a 'feminized household,' not necessarily of her own will. She had had to raise five children in the absence of a husband who chose to live away for more than thirty years, visiting her occasionally (sometimes as rarely as twice a year). In the meantime, she and her eldest daughter Hana practically held the family together. Although her daughters quit school at an early age, she encouraged them, especially Hana who never married, to build a career, to be independent, and to safeguard their own future. She was proud of the fact that Hana travelled the world by herself, even when commentators protested that 'This was not the Arsali way.' Um Yusif contested the cultural tendency to have lives that were only cast in the shadow of male kin (Keddie and Baron 1993; Kandyoti 1995; Joseph and Slyomovics 2001; Delaney 1995) and sought to regard and treat her sons and daughters equally. In this comparison, the dystopian face of 'anti-kinship' loomed large: at the end of the day, Um Yusif's mother repeated the cultural preference for sons, her brothers failed her at several times of need in her life, and her husband left her to cope with raising a family on her own. If this was meant to be some sort of 'patriarchal bargain,' then someone was not keeping their side of the deal. Um Yusif ended up with a failed bargain.

3   The Morality of Kinship

Um Yusif's conflict with her brothers elaborates the other face of kinship (anti-kinship), where positive attributes of intimacy, affection, and trust are replaced by selfishness, betrayal, and pain. This tension between the expectations of kinship and the contradictory experiences of it left Um Yusif in a morally ambivalent situation. She had no doubt that she was in the right. Her brother had hurt her son and disregarded her feelings. Her mother had sided with 'the oppressor' (al-*dhālim*), privileging her sons over her daughters, as she had done before in Um Yusif's lifetime. Yet 'dis-acknowledging' her family by isolating herself from them somehow did not feel like the right thing to do. She believed that this was not how kin ought to behave towards each other. In the weeks that followed, as she pondered on the unjust nature of her relationships, Um Yusif was subjected to the power and morality of kinship. Although her audience was empathetic, the responses and messages she received from her sons, daughters, neighbours, and friends pressed her to honour the sanctity of kinship and reminded her of kinship's 'diffuse, enduring solidarity' (Schneider 1968). This power of kinship was derived not only from the morality inherent in kinship as an ideology but also from the way Um Yusif's interlocutors drew on Islam to consolidate the importance of kinship. In this sense, we may think of kinship and Islam as two indices of morality, as moral 'registers' (Schielke 2009) that draw on each other and sometimes converge.

'Did you go to see your mother today?' Hana asked Um Yusif one late morning as she anticipated her response. Tuesdays were the days when Um Yusif visited her mother, bathed her and asked for her blessings. 'I cannot get myself to see anyone,' Um Yusif responded sadly with a hint of guilt. '*A-hoo* mother! How long will it be? *Rida Allah wa rida al wālidayn!*' Hana reminded her mother. She was referring to the well-known Qur'anic verse, 'Show gratitude to me and to thy parents' (Chapter 31, Verse 14). Through this verse, the Islamic ethic elevates parents by making their blessings as important as God's. Denying parents is therefore akin to an act of sacrilege. The Arsalis disapproved of people who neglected their elderly parents. For example, in another neighbourhood, Um Khalid was appalled to see one of her neighbours, an elderly man with possible dementia, sitting on the ground at their street corner, begging for food and reeking of urine. He lived alone in his house after his married sons moved to different parts of Arsal. He would often forget where he lived and wandered aimlessly around the streets until a kindly passer-by recognized him and walked him back home. 'God will not have mercy on his children,' Um Khalid assured us. '*Rida Allah wa rida al wālidayn!*' She blamed his sons for not honouring the respect of and care for kin that God demands.

The verse also implicitly places the mother on an equal standing with the father. This resonates with the Arsalis' idolization of motherhood. They often described the mother as *um al-dunya* (Mother of the Earth), glorifying motherhood beyond the realm of the family, as a virtuous role that affects life in its entirety. Mothers were talked about as symbols of sacrifice and giving (*'atā'*) and as emblems of female altruism. This made Um Yusif's critique all the more difficult as she had loved and respected her mother and had conscientiously carried out her obligations toward her. Um Yusif expected her mother to show some kind of recognition of her dutiful daughter. Her anticipation of an intervention, mediation, or attempt at a family reconciliation following the fall-out over the factory was thwarted when her mother did not react to her daughter's deliberate withdrawal. It was then that Um Yusif was torn between guilt about not acting as a proper daughter and anger towards her mother and brothers for neglecting their kin obligations. She found herself in a situation of 'moral breakdown,' when an 'ethical moment' arises and makes ethical demands on a person (Zigon 2007: 138). Um Yusif's moral and religious sensibilities eventually trumped her anger.

About a month or so later, Um Yusif extended an olive branch by going to her mother's house. Hana and I waited for her, hoping that this visit would finally unburden Um Yusif. Upon her return, she reported that her mother had acted as if nothing had happened. This took Um Yusif aback. She did not confront her mother. Instead, she refused to accept any food, a rejection of commensality and a statement that their relationship was still unsettled. Um Yusif told us that she broke into a flood of tears but no one was moved to ask about the cause of her misery. Being an expert in the language of silence, though, Um Yusif felt that her kin knew what they had done, and were aware of all the pain they had caused. The factory fall-out thus ended without confrontation but also without compensation. To Um Yusif, her son's rights counted for nothing and so, once again, did hers. On the occasion of *Īd al Fitr*, the feast that falls after Ramadan, her brother Fahd paid a visit to her house. He was offered sweets and coffee as he chatted with her and her children. His gesture and Um Yusif's hospitality in return were a statement that the problem was over. Reconciliation had taken place albeit without recognition of the injustice. As Jarrett Zigon (2007: 138) suggests, the crucial demand in ethical moments is to get out of the breakdown, to 'keep going.' 'God sees,' Um Yusif concluded, resigned to the idea that in the end only God grants justice. Hence it was best to move on, especially since 'God has asked us not to cut the relations of the womb (*qatʿ silat al rahm*),'[7] she continued. This Qur'anic reference also suggests that maintaining kin relations, particularly those between close kin, is a religious duty. It was often invoked in contexts that needed an assertion of the long-term nature of kinship and the virtues of enmeshment and connectivity. Its aim was to confirm that kin could not, and ought not, undo their relationships. To do so would be unethical by social and religious standards, for abandoning kin was considered a wretched and deplorable thing. Perhaps this explains the reluctance of Um Yusif and her son to admit, in the vignette at the beginning of this chapter, that what they were describing was a *khilāf* (disagreement, conflict), a term they avoided in the few weeks after Yusif's dismissal. *Khilāf* might have implied that the relations of the womb were broken. In the next chapter, I look at the ambivalence of kinship through another lens. How does kinship, in its loyalty and treachery, play out in local politics? This question is central to a political system that is entrenched in kinship.

---

7 See Morgan Clarke's discussion of *nasab* in Islam (2009: 95).

CHAPTER 7

Local Elections: Politics at the Margin

Um Abdu's daughter was busy brewing two large pots of tea and coffee as her sister prepared plates of fruits and nuts for Abu Abdu's guests. There were some ten men in the living room, passionately engaged in a discussion that was turning noisy as the room filled with their cigarette smoke. From the kitchen, we could also hear Um Abdu and two other women interjecting comments as the guests talked over each other. The voices were getting louder and Um Abdu's daughters sensed my curiosity. I had just arrived at their house and had gone straight to the kitchen to greet the daughters. '*Intikhābāt!*' ('It's the elections!'), the youngest (aged 15) commented with a sigh. 'When else would you have the pleasure of hearing this symphony?' It transpired that in the past week lobbying had begun for the upcoming 2004 local elections. The daughters' evenings had been sabotaged as talk of candidates, political parties, past mistakes, and future needs swamped all other conversation. '*Ya 'abī* (oh father!), they say women are talkative! These men can talk your ear off! And my mother, can you hear her? She has become quite the philosopher!' We giggled as the older sister (aged 23) teased the younger one about putting her dubbed Turkish soap opera above the future of her town. Wasn't she really irked because she had missed that evening's episode? Carrying a hospitality tray each, the three of us joined the visitors in the main room.

'Putting a bunch of candidates together in an electoral list based on their *'ā'ila* (lineage, pl. *ā'ilāt*) alone won't do it this time!' Abu Abdu was saying. '*Khalas!* (Enough!) It is time that we put the needs of Arsal above our *'asabiyya 'ā'iliyya* (familism or extreme loyalty towards kin).'[1] Abu Khalid seemed to agree but was not fully convinced that lineage loyalty could so easily be ignored. 'We are a familist society (*mujtama' 'ā'ilī*) in the end. You cannot ask the Arsali to vote for someone who is not from his lineage. That's how we are.' Another man went further, 'The Arsali will always vote for his cousin, even if his cousin were the *Shaytān* (Devil) himself!' The guests laughed at this suggestion, but he continued gravely: 'The 1998 [local elections] worked *only* because

---

1 The concept of *'asabiyya* was coined by Ibn Khaldun who was 'referring to tribal groups, for whom *'asabiyya* has a specific meaning, that is the feeling of solidarity or cohesion among the members of a group that is derived from the knowledge that they share a common descent' (Alatas 2006: 401). In common parlance, this term indicates a more exaggerated meaning of rigid and extreme forms of loyalty.

the lineages were equally represented. Had this not been the case, we all know that we would have had a catastrophe; 1964 all over again.' He was referring to the year that followed the 1963 local election, the last to be held before the 1975–1990 war. Considered a dark moment in Arsali history, violence broke out after the results because members of a particular lineage felt cheated out of their entitlement to a larger representation. The repercussions of these events reverberated for years to come and the wounds and rifts they caused lingered on after the war.

'*La, la!*' – a succession of multiple-pitched 'Nos' resounded through the room as his listeners rejected the idea that in the here and now people would turn violent if their candidate were not chosen. Um Abdu reassuringly stated that 'Today, there is *waʻy* (awareness). People are no longer like they used to be in the old days,' the implication being that the cause of the violence lay in people's (past) nature rather than the external circumstances that led to its outbreak. Abu Abdu backed his wife and reminded his audience that 'humans evolve' and that the previous 1998 election was 'actually exemplary. Didn't the Arsalis respect democracy?' He argued that the success of that election and the triumph of democracy were proof that

> *al-yawm* (today) the Arsali follows his *aql* (reason) rather than his heart. Let us leave the *qadīm* (old days). We learn from [the past], but today we are different human beings. We need to find the right candidates, with decent reputations and an ability to represent us. The stakes are too high now.

Over the next three weeks, the themes explored in Um Abdu's living room were endlessly rehearsed at every social gathering, whether in houses, shops, on public buses, or at NGOs. The residents of Arsal pondered over what political actors were needed in the town's post-war context. As in the 1998 election, the relationship between kinship and politics took centre stage in these discussions. Was it possible for voters to reconcile the primacy of lineage as an idiom of representation with democracy, especially given the town's past experience with familism and its consequences? What lessons could be drawn from the past two elections? While 1998 had been successful as an electoral exercise, as Abu Abdu maintained, the elected council had failed to address the town's most urgent needs. Was this a result of misguided voting? And, bearing in mind the overbearing interventions of the Syrian state, which – particularly since 1998 – had increasingly used its apparatuses to challenge the sovereignty of the local authority in Arsal, what kind of competency would a candidate for the council be required to have? Given the novelty of democratic elections in Arsal, where only three municipal elections had been held between Independence in 1946 and the present one in 2004, my interlocutors approached the electoral exercise as a course of experimentation (*tajruba*), as something that needed working out – in the words of Ali, a local political activist, a sort of 'trial and error' in politics – rather as they did with their livelihoods (Chapter Three). Voters' views, discourses, and choices in this election were informed by their cumulative experiences of previous candidates and councils. 'Temporal consciousness' was therefore at the heart of the political processes described in this chapter.

In what follows, I focus on local elections, also known in Lebanon as municipal elections (*intikhābāt baladiyya*), as a site for exploring shifting political relatedness and affiliations in post-war Arsal. The shared cross-cultural language of 'elections,' like 'politics' and 'representation,' can render these social practices 'culturally-neutral' (Coles 2004: 552), thus obscuring their local specificity and ambivalences. A focus on the ethnographic, therefore, allows us to explore different 'areas of life which are implicated in what people take to be politics' (Spencer 1997: 12). *ʿĀʾila* is a central political idiom in Lebanese politics and is a 'more common referent for the Lebanese … across religious sects … than their national identity' (Joseph 1999: 298). Kin and sect loyalties, which are often interlaced, instigate ambiguous and contradictory discourses in Lebanon. These institutions (their moralities, idioms, and affects) are enmeshed in the Lebanese political fabric and they often stand in for the state. A person is more likely to resort to a sectarian *zaʿīm* (leader) to secure public service employment than go through official state channels. A cousin working for a ministry might speed up an application for a permit, by-pass bureaucracies, and 'talk to the right people.' For this reason, kin and sect are regarded as traditionalist primordial forces that reproduce patron-client relations and contribute to the reduction of the state and the reproduction of unequal citizenship (Khalaf 1977; Hamzeh 2001). It is not uncommon for the same people who mobilize nepotistic relations to advance their agendas to critique the 'backwardness' of these same practices as disrupting modernity. During democratic elections, discourses about the benefits and vices of affiliations through *ʿāʾila* and sect become all the more rife. In his historical survey of municipal elections in Lebanon, Muhammad Murad concludes that municipalities have remained 'prisoner to a traditional structure of strong lineage loyalty (*ʿasabiyya ʿāʾiliyya*)'[2] which the author considers 'backward (*mutakhallifa*)' (2004: 43) because it hinders the democratic process of governance. Through the material

---

2   Murad's survey of municipalities in Lebanon since 1963 shows that 78.9% of elected mayors belonged to the five largest lineages and 44.7% belonged to the largest lineage of their village,

in this chapter, I take a different view that sheds light on the democratic potential of *'ā'ila* as a political idiom.

The following sections discuss and compare three consecutive municipal elections (1964, 1998, and 2004) in Arsal. A comparative approach within the same setting will bring out the transformation in local discourses about lineage and its vital role in the administration of the town. The idiom of *'ā'ila* is malleable and is shaped and reshaped by the socio-political environment in which it is embedded so that it is possible for it to be unifying or divisive, or even a principal idiom of democracy.

1   1963: Familism, a Divisive Force

In one of a series of conversations I had with an eloquent schoolmaster, we began to talk about kinship. I told him that I was interested in researching changing kinship relations in Arsal, using the Arabic word *qarāba*. I had not expected his strong reaction: "*'ā'iliyya* (familism) in Arsal is the greatest of all catastrophes (*musība*)!' he declared. To make his case, he raised the infamous date of 1964: 'Because of *'ā'ila*, Arsalis ended up slaughtering the sons of their own town!' What exactly happened that year was difficult to establish. I researched a number of Lebanese newspapers[3] but could find no coverage of the events referred to by Arsalis when they mention that year. In fact, the only clippings relating to Arsal in 1964 reported killings, robberies, and clashes with the authorities, the kind of stories that reproduced the historical stereotype of the Northern Biqaʿ as a frontier that existed outside the law. However, in the course of my interviews and informal conversations I encountered more than one version of the story, depending on whose side the narrator was on. Nevertheless, all the versions I heard had certain core events in common.

My interlocutors maintained that the problems arose when a challenge was made to the leadership of one of the largest lineages. 'For more than 200 years the *zaʿāma* (leadership) of the village had belonged to the Jawhar lineage,' Abu Ali explained. But in the 1963 municipal election, the smaller lineages allied together against the Jawhar candidate and voted for a mayor and a vice-mayor from among themselves. This angered members of the Jawhar lineage

---

town or city (2004: 99). He sees these results as an indicator of the importance of *'ā'ila* in municipality leadership.

3   With the help of a research assistant, I surveyed the following dailies: *Annahar*, *Assafir*, the Lebanese Communist Party's *Al-Nidā'*, the Progressive Socialist Party's *Al-Anbā'* and the Nasserite *Al-Anwār*.

who believed that they were entitled to representation on the council. Losing council seats jeopardized their control over the local authority (*sulta mahalliya*) and, inevitably, the management of local resources including municipal funds, taxes, and fees (Hamzeh 2000: 744). They therefore refused to accept the election results and pushed to dissolve the council.

In the months that followed, troublemakers from the Jawhar lineage began to stir up violence, harassing their opponents, and even throwing stones at them. Their anger was directed particularly at members of the Karama lineage, who they believed had masterminded the alliance against their candidates. The scheming was described to me as a '*tabkha*' (stew), a food-related word that suggested the row had been cooked up slowly and over time. Although the town witnessed several clashes in the aftermath of the election, one in particular was widely identified as the trigger for the eruption of violence. A member of the Jawhar lineage was said to have stood outside the house of one of the Karama, shouting threats and insults at his family and female kin. 'The offense was too big!' Abu Ahmad told me. 'They insulted his honour ('*ard*)' so much that the man from Karama 'lost his mind' and shot and killed his abuser. The Jawhar retaliation was brutal and the bloodshed that followed spiralled out of control. I was told that a significant number of terrified Karama families fled, some of them never to return, relocating permanently to another village in the Biqa' by moving their *nufūs* (official documentation) there.

This brief summary of what happened in 1964 cannot fully convey the way the events were differently inflected in detail and emphasis depending on which '*ā'ila* the narrator sympathized with. The divergences suggested competing 'truths.' The Jawhars I spoke with, for instance, highlighted the way the other lineages had betrayed them by ripping them off their fundamental right as the majority. For them the election results were an act of *zulm* (injustice). To the Karamas, on the other hand, the events confirmed that the Jawhars had never intended to share power and resources and were testament to their 'violent nature,' the manifestations of which, they felt, were still palpable at the present time: 'Where there is a fight in the town, it will be a Jawhar. This is their nature (*tab'uhum*),' one man complained. Other Karamas suspected that members of yet another large lineage had tricked them into being made the scapegoat and abandoned them after the killing to face the rage of the Jawhars alone.

My discussions with various people revealed further dynamics of familism. Although the most common story described a rivalry between the Jawhar and other lineages, other accounts indicated that there were growing tensions within the Jawhar lineage itself, as Abu Hussein, a Jawhar himself, explained to me. He speculated that members of one branch could have been supporting the smaller lineages in order to break the other's traditional power. It was only when a Jawhar failed to be elected as mayor that both factions realized that their lineage had been isolated and resolved to unite in the face of the threat. Abu Hussein contended that the Jawhars used their political connections beyond the town to obtain official support for dissolving the council. 'In those days, the small lineages knew nothing outside of the borders of Arsal. But the Jawhars, you are talking about a *za'āma* (leadership) that has influence not only in Arsal but in the entire Biqa' area.' It was understood that the Jawhars used their connections with the governor (*Muhāfiz*) of the Biqa' to issue a decree in 1967 dissolving the municipality council after a series of 'inspections.' This type of manipulation was not uncommon; *zu'ama* (leaders) in Lebanon frequently benefited from relationships of patronage and clientalism to reap exceptional powers (Gilsenan 1996; Johnson 2001; Jabbra and Jabbra 1978). Whereas some Arsalis claimed that the inspections had found nothing, others contended that the mayor was reported to have embezzled public resources.[4] Abu Hussein pointed out that the sons of the two rival Jawhar leaders in the 1960s were elected as Members of Parliament in 1996 and 2000 respectively. This, he felt, proved Jawhar's connections to the state and centres of power, as it was inconceivable that they could have attained these posts without associating themselves with weighty political figures within central government who already enjoyed mass support.

At one level, these events were understood as the culmination of local tensions driven by familism, which had played on people's passions in unprecedented ways. But beyond this purely local scenario, some of my interlocutors proposed an analysis that extended outside the boundaries of the town and drew on the historically volatile relationship between the town and the state. A local researcher, for example, believed that the *ajhiza*[5] (apparatuses; namely the secret service) had a particular agenda in Arsal and a vested interest in the escalation of conflict that led to violence.

---

4 I heard further accusations of conspiracy in these discussions. One man claimed that certain agents or 'parties' (*jihāt*) encouraged the mayor to steal money, persuading him that it was his right. The man claimed that it was that same party that later reported the mayor to the authorities.
5 During the years of my research, the term *ashbāh* (ghosts) was used by the media to suggest that the secret police stirred events in an invisible manner. This was particularly so during the regime of President Emile Lahhud (1998–2007), who chose to employ military advisors to manage Lebanese internal affairs and strengthened the apparatus of the secret police.

> The *ajhiza* of the Shihabi[6] rule [secret service of President Shihab] played a role ... it has come to my attention that approximately 100,000 Lebanese pounds were paid [as bribes] to people inside the council and outside it. [They were expected to] throw stones at this person or stir trouble between two houses. The result was that problems escalated and became complex. The state found an opportunity to bring the Army in and to discipline Arsal.

Why was the state interested in subduing Arsal and its people? This researcher and others I spoke to linked the violence of 1964 to a set of events that had taken place in the 1958 civil war, when an Arab nationalist-led rebellion against President Chamun, who had controversially endorsed the Eisenhower Doctrine, seen as consolidating his affiliations with the West, broke out across the country. The Arsalis took part in the protests, leading to skirmishes with neighbouring villagers who were supporters of the state. An elderly interlocutor who had taken part in the fighting told me that the Arsalis defeated their neighbours, thus proving themselves to be a power in the Northern Biqaʿ: 'We made it known that Arsalis were strong people who would not be bullied.' He suggested that the emergence of these 'strong' people was a concern for 'the state,' which looked for an excuse to tame (*taʾdīb*) them once the rebellion had ended. Events following the 1963 election provided a credible opportunity, especially after the townsmen started fighting the security forces deployed to quell the violence. The situation escalated to such an extent that, around September 1964, the army besieged the town (*Annahar*, 27.9.1964: 4). The national daily newspaper, *Annahar*, published a powerful letter from residents pleading with the President of Lebanon to intervene to end the violence.

> Arsal weeps for the blood of its children ... this village that lies on the borders of Lebanon has been treated today as if it is a village in Israel, as if it has no pride or nationalism. We call on your ... nationalism and beg you to take the necessary measures to move the weapons pointed at our chests, we the children of Lebanon, and point them at the chests of the enemy [Israel].... We celebrate your rule [and] ... have true national feelings: we are Lebanese and have the same rights as all other Lebanese. (ibid.)

---

6 Fuad Shihab, President of Lebanon in 1964, famously strengthened intelligence and secret services during his rule, better known as the *shuʿba thāniya* (the second branch) or *maktab thānī* (the second office).

This appeal would resonate even more strongly in the post-war years, when Arsalis began to contest their predicament of living under the military rule of the Syrian state by asserting their Lebanese nationalism.

The repercussions of the 1963 election brought home the underlying threatening nature of *ā'ila* and its capacity to act as a divisive force in society. It is to the 1998 municipal elections that I now turn. My aim is to trace how people generally sought to avoid the failures of the pre-war election by pushing for the consideration of new criteria for leadership that promised change but had not yet been put to the test.

2   1998: *Ā'ila* Redeemed

The first post-war local election was a big event across the country[7] that indicated a 'return' to 'normal' political and administrative governance. Given Arsal's last election, the event seemed even bigger there. Newspaper and television reporters flooded the town during the week of the election. For the first time since the end of the war, an army regiment was sent to Arsal during the election period to assist the force of twelve gendarmes (the number had been increased from two in 1995) in charge of the town's security. A heightened sense of anticipation loomed over the town: would 1964 repeat itself? Would there be violence? How were the lineages going to act and react? But the election went ahead peacefully and calmly, so much so that Um Abdu felt it was 'as if you were in Switzerland!' Like others, she was impressed with how *hadāriyya* (civilized) the whole process was.

The town had started buzzing weeks before that as the electoral lists (*lā'iha*, pl. *lawā'ih*) were put together. Voters could vote for up to 21 individual candidates. The lists contained candidates who promoted a common programme or political affiliation, the aim being that voters would opt to vote for a bloc list of nominees rather than individuals from rival lists. Lists were also given names that represented their outlook, for example 'the Democratic Choice.' These lists were photocopied and distributed to voters. On election day, a voter could either post a full list of names in the ballot-box or strike out the names of unwanted candidates and add new ones. Once elected, the council would

---

7 In 1997 a national civil society campaign, '*Baladī, Baldatī, Baladiyyatī*' (My Country, My Town, My Municipality), gathered thousands of signatures to hold new elections and to reinstate the municipalities that had been dissolved before the war. Muhammad Murad (2004) suggests that President Ilyas Hrawi (1989–1998) supported the campaign because he wished to end his rule with a success.

vote for the mayor and vice-mayor. Since their period of office was six years, the council was given the right to change both the mayor and vice-mayor after the first three years of office.[8]

After months of enthusiastic politicking, three electoral lists were constructed. All three had candidates from different lineages. One in particular promoted itself precisely through the idea of fair proportional representation of the different *ā'ilāt* (as opposed to a random presence of different lineages) and was even referred to as the Lineage List, despite its official name *Al-Wihda Wa al-Ta'āwun* (Unity and Cooperation). The effort made to highlight this configuration perhaps reflected the general nervousness arising from the town's history and represented a pronounced attempt to avoid lineage-related conflict. As for the other two lists, their formation focused more on their political affiliation – or lack of it. The *Al-Khayār al-Dimuqrātī* (Democratic Choice) consisted of 16 candidates representing the Communist Party. Al-Tawāfuq (Consent) had eight independent members, who were neither affiliated to any political group nor advertised themselves in terms of lineage. I will refer to the first two lists as the Lineage List and the Communist List, as the locals did.

The results of the election disappointed few of the people I knew. Sixteen out of 18 candidates from the Lineage List were successful, with only five elected from the other two lists. The fact that the town's residents had been able to construct a list perceived as representative of its various lineages was an achievement in its own right. That this list won by a majority was cause for celebration as it suggested a mood of *wifāq* (accord) in the post-war years, as many people reiterated to me when discussing the results. However, for those like Zuhayr (Chapters Two and Five), who maintained their political ideological affiliations, the results were only the beginning of a series of disappointments. Zuhayr and his Communist Party comrades felt a sense of failure that '*ā'iliyya* was still the overpowering concern for the Arsalis,' when people like them wished to embrace a broader nationalism that transcended considerations of family and clan. The members of the new council, he worried, 'would think Jawhar and Karama and this *ā'ila* and that! But a Communist would think Arsal and Lebanon.' Although the Communist Party had long been established in the town, its popularity declined considerably after the 1975–1990 war. Mocking the 'emptiness of their ideology' (see El-Khazen 2003), opponents of the Communist List teased Zuhayr mercilessly for months after

---

8 In December 1997, a new Municipal Law (665) was introduced. Prior to that, the mayor, who had decision-making as well as executive powers, had been elected through a direct vote by the people. The new law sought to limit such power by introducing measures of accountability and reducing familist-style patronage.

the elections, referring to the party list as the Democratic Cucumber, because *khayār* (choice) is phonetically similar to *khiyār* (cucumber). The privileging of lineage over party in the 1998 election was not particular to Arsal. It was observed in other areas of Lebanon as well, and was ascribed to the general failure of political parties in the post-war period (Hamzeh 1999; El-Khazen 2003).

The 1998 electoral exercise seemed to redeem *ʿāʾila*, which was now reified as a vector for unity and fairness in the town. In addition, the election results were seen to alter the local political dynamics prevalent in the 1960s, which favoured a single powerful *zaʿīm* (leader) from one *ʿāʾila*. Instead, the victorious list managed to include weighty, qualified figures both within and across lineages: a young lawyer, a medical doctor, and a civil society activist who had longstanding experience in political parties and NGOs. The hopes for this council were raised further when, to allay people's fears over the lingering tension over who (which lineage) would assume the office of mayor, the council agreed to split it into two terms: the first held by a political activist from the Karama lineage and the second by a medical doctor from the Jawhar lineage who was highly regarded in the town. The Lineage List had to that point lived up to its name of Unity and Cooperation.

The 1998 elections thus ended with a sense of achievement: the weight and power of *ʿāʾila* were asserted, the big lineages were well represented, and those in power had the right qualifications – education, experience, and social status. There was general satisfaction that the residents had been able to overcome the divisiveness that *ʿāʾila* had previously brought to their town. As soon as the celebratory mood passed, however, the council was faced with the real challenges of administering the town with a budget that could hardly meet the accumulated list of urgent needs bleeding over from the war period: from building roads to upgrading sewage networks to regulating land use, among many others. People expected to see some change and began to scrutinize the council not only in terms of its formation but also its performance (*ādāʾ*). The developments below bring to light the lessons that emerged out of the 1998–2004 municipality council experience. They also illustrate the malleability of *ʿāʾila* as an idiom that is entrenched in the larger social and political environment.

3   Familism Strikes Back

A few weeks after the election, the uniting nature of *ʿāʾila* was soon put to the test as inter-lineage tensions resurfaced. Municipality public works suffered deliberate acts of destruction (*takhrīb*). Some thugs, people complained, were damaging and breaking streetlights. Matters became more serious when dynamite was planted near the mayor's house. The Karamas, to whom the mayor belonged, accused the Jawhars, and the voices and images of 1963 resurfaced. But the mayor ignored these provocations by opting for what some considered to be a 'pacifist approach.' '1963 was a *'ibra* (moral lesson) and the mayor must have felt that it was better to stay quiet than to repeat history,' explained one of his supporters. According to this man, the mayor instructed members of his lineage not to retaliate. Others, however, felt that the mayor was more of a strategist than a pacifist. Halfway through his term of office, he was believed to have secured two main sources of support, both deriving from apparatuses of force belonging to the Lebanese and Syrian states. He was credited with officially requesting the deployment of the Lebanese army by apparently using an appropriate opportunity, a fight that escalated between two young men from rival lineages, to make a case for further security. For the first time, an army base was set up permanently inside the town. Although acts of vandalism were kept under control after that, my interlocutors complained about the army's unnecessary use of intimidation to subdue the residents (Obeid 2010b).

The second source of support was more momentous in the wider context of the Biqaʿ area. The mayor, I was told, allied himself with the notorious Syrian intelligence, the *mukhabarāt*. By the late 1990s, the political climate throughout Lebanon was becoming highly charged. The Syrian state was believed to loom so large over Lebanon's political life that it controlled the appointment of cabinet ministers and Lebanese presidents. Syrian intelligence officers were known to manipulate and make decisions about the composition of electoral lists for parliamentary and municipal elections across the country (Salloukh 2005). The blatancy of Syrian control was particularly visible in the Biqaʿ region, not least because the Syrian intelligence headquarters were based in the Biqaʿi town of ʿAnjar, which 'became the locus of decision-making in Lebanon and symbolic of Syrian control' (ibid.). The anti-Syrian sentiments that were brewing in Arsal in the 1990s echoed those in the country at large, with the difference that Syrian control in areas of the Biqaʿ like Arsal, whose politics were not 'pro-Syrian,' was less subtle and affected people's lives more directly than in the capital. For example, Syrian intelligence officers blatantly used violence to intimidate and humiliate citizens. They made unlawful arrests in Lebanon and detained those arrested as prisoners in Syria. The Lebanese could not get their kin released from Syrian prisons without first handing over a significant bribe to find out where they were being held, and then paying more to get them released. People understood these episodes to be part of an economy of bribes through which Syrians profited at the expense of the Lebanese.

Like other Lebanese, Arsalis complained about the common Syrian practice of extracting bribes from commuters, businesses, and public and private companies.[9] My Arsali interlocutors described all Syrian officers, regardless of their rank, as 'petty.' 'They can't live without *tazyīt*! Like a car, they need lubricating,' Ahmad told me scornfully as he explained the mechanisms of bribery. Disparagingly, he believed the most senior Syrian officer 'could be bought with a pack of bread.' His friend Imad told us that once when he was transporting a load of watermelons, as soon as 'the officer saw the watermelon in my hand, all his power melted away.' During the inspection, Imad offered him the large fruit as a 'gift' from him to the officer's family. The officer, who was expecting a 'gift,' accepted it and let him go. While the scorn with which these stories were related may have granted their narrators some sense of dignity by revealing the pettiness of the Syrian officers involved, it also betrayed their own helplessness, especially in light of the potential risks of harm and humiliation had they tried to challenge an officer, regardless of his rank. These tensions were sometimes expressed in a humorous manner that perhaps attempted to downplay embarrassment and disgrace. For example, at an evening gathering, a senior member of the council recounted how he once attended a wedding, where he happened to sit next to a known *mukhabarāt* officer. He had been disabled by severe back pain for some time: 'I had not danced for more than 15 years! For a whole month before this incident, I had lain on my back on the floor, as I couldn't move. *Wallahī* (I swear), if my wife had asked me to dance at that moment, I would have divorced her.' But at the wedding, it was the Syrian officer who asked him to dance the *dabka*. He complied by placing himself in a line of dancers between the officer and a small woman. Laughing aloud with his listeners, he described how he had to twist his body and lean his full weight on the poor woman who 'must have felt like she was carrying a barrel of milk,' just so the officer wouldn't think he was disobeying his request. 'The *mukhabarāt* says "dance!" you dance, even if it breaks your back.' Stories like these acknowledged that Syrian power was a fait accompli. 'Nothing could be done with the Syrians!' This sigh of resignation echoed across the country.

In this climate of opinion, and amid the anti-Syrian politics developing in Arsal, the mayor's decision to collaborate with the Syrians was cause for alarm. A council member, who happened to be a relative of the mayor, believed he gave in to Syrian pressure when he broke the post-election agreement about handing on the second term of the mayoral office to the Jawhar

---

9 Salloukh (2005) refers to a study published in the *Annahar* newspaper (March 25, 2005) that estimated the revenues generated by checkpoints across the country from 1976 to 1990 at around $1.6 billion USD. The study also argued that between 1976 and 2004, private and public companies paid Syrian intelligence officers fees estimated at around $5.4 billion USD. Moreover, the Syrian military and intelligence presence in Lebanon between 1976 and 2005 is said to have cost Lebanon an estimate of $27 billion USD.

medical doctor. He believed that the Syrians wanted their 'own man,' who was also a Jawhar. The mayor is said to have engineered the vote in his favour but, unlike the doctor, the new mayor had not been educated to a high level, nor had any previous experience of public life. When he was appointed, several of my interlocutors complained that the council had picked an 'illiterate man.' This called into question the new mayor's ability to speak for the town and confirmed to them that he was merely a 'Syrian puppet' put in place to carry out the agendas of the Syrians. At another level, members of the Jawhar lineage felt betrayed by the former mayor, since as a fellow Jawhar he had overtly broken the bonds of *'ā'ila* – in stark distinction to 1963 when two factions of the Jawhar were believed to have eventually united against an outside threat. More worrying to the mass of people was the fear that Syrian intelligence officers would now control the town's resources, negating the principles of *hukm mahallī* (local governance) and sovereignty. In the event, the council's failure to carry out even minor infrastructural works was attributed to the mediocrity of its leadership in combination with its subservient attitude towards Syrian intelligence officers, who were seen to indulge in acts of corruption (*fasād*) that only entrenched Syrian control. The following events, which took place a few months before the 2004 election, particularly mobilized people against Syrian state dominance and prompted them to rethink their voting strategies in that election.

4   Corruption that Compromises National Pride

The controversy began with the circulation of second-hand accounts of an auspicious dinner that had taken place at the municipality building. The guests included council members, *mukhtārs* and many other key figures, including a Syrian *mukhabarāt* officer, and the meeting was said to have resulted in the decision to build a permanent checkpoint at the entrance to the town. Paranoia spread, as residents feared that Syrian intelligence agents would control the movements of Arsalis in unprecedented ways. While '*relatively* autonomous of corrupt actions and [while they have] their own life and efficacy,' rumours and narratives of corruption 'do not function independently of such actions' (Gupta 2005: 18; original emphasis). The rumours were confirmed when construction work soon became visible halfway up the hill that leads to the town. The Syrians *were* going to control the mobility of the locals. 'But can they do that?' I asked Hasan, a truck driver who had stopped by the NGO and was offered a cup of tea as we discussed this issue. I wondered whether there had been a misunderstanding.

They can do anything they want, who can stop them? *Bi Allah* (by God), my friends and I feel like giving the mayor a beating. He has always been a *mukhbir* (spy) for the Syrians. This is his doing.

Syrian checkpoints were usually posted on main roads, not at the entrance to villages or towns. A couple of the women and I pondered over how the Lebanese government might justify this move, given that there was an army unit stationed in the town already. Najwa had heard talk of a terrorist group that the Syrian and Lebanese governments were after. But Hasan poured scorn on this excuse, which he too had heard. He explained it was merely a smokescreen. The group referred to in these accounts, *Al Takfir Wa al-Hijra*,[10] did exist, but its insurgents operated in the northwest, not the northeast, of Lebanon. He felt that the official version was using the *jurds* (highlands), an area that could potentially harbour outlaws, as justification for a cover-story about terrorism. In his version, the Syrian *mukhabarāt* only needed an 'invitation' or endorsement to aid the local council in the name of national security. The mayor would have had to be the official party to sanction this level of security. The real motives, he suggested, were obvious. A permanent checkpoint would allow Syrian officers a systemic channel for bribes.

A quarry-worker later added details to this perspective on events. He believed that the decline of smuggling by the late 1990s must have incurred losses for the Syrian officers who, when smuggling was at its height, had relied on a stream of bribes in return for turning a blind eye to illegal operations. The timing of the checkpoint was of relevance, he suggested, because it came at the same time as the doubtful quarry law already mentioned in Chapter Three. The quarries were ordered to cease extraction, but were given a grace period to sell their stock. This was seen to be a perfect opportunity for the Syrians to exploit the situation and demand bribes from truck drivers on their way out of the town. If they refused, the Syrians would not allow them to leave Arsal. This conspiratorial line of thought represented the council, led by the pro-Syrian mayor, as acting in the interests of the Syrians at the expense of the local population. The performance of the council was considered a form of *tabyīd wajh* (lit. whitening faces; currying favour) that aimed to placate the Syrian apparatuses in the hope of getting the Syrian officers to loosen their grip on the council. This was seen as a two-way deal, especially as council members were suspected of benefitting from the suspension of quarrying activities. During

---

10   This group is believed to be a subsidiary of the Muslim Brothers. *Takfir* means excommunication, while *hijra* refers to the pilgrimage that every Muslim is required to undertake.

this period, suspicions were fed by accounts that pointed to the council's exploiting the ambiguity of the quarrying laws.

As people worried about how they would make a living now the quarries were to shut, Hussein, a quarry owner, and others in the quarrying industry heard that the council had taken measures to impose a new tax that, upon payment, would allow the quarries to resume operations. This was encouraging, as it gave people hope that the industry was going to be regulated rather than shut down altogether. But Hussein and his co-workers subsequently learned that the tax had nothing to do with guaranteeing the regulated continuity of quarrying activities. As mentioned above, the quarries had been given a three-month period to dispose of their stock. What angered Hussein was that they only learned about this as result of a visit from a Lebanese officer (*muqaddam*) who had been sent to ensure that Arsal was abiding by the new law. The tax imposed by the council, he told me angrily, 'had nothing to do with the government but rather went in to the mayor's pockets!' If the council had a rationale for this tax, it was not clear to many of my interlocutors, who had come to mistrust their representatives. Quite apart from its dubious integrity, Hussein felt that the council was incapable of explaining the town's situation to the central Lebanese government.

> It was so shameful when the Lebanese *muqaddam* was here. This municipality [council] ... not one of them knows how to speak. The mayor ... he knows nothing. I raised my hand and told the *muqaddam* that now that he's seen Arsal – and he was shocked! – he should go back and tell the government that Arsalis work with rocks but can't [even afford to] use them to cover their own houses. Everybody thinks that Arsal is the richest village because of rocks. I asked him to tell the truth to the government so maybe they can be merciful with us. But the mayor, he interrupted me and said, 'Save your words and don't give the *muqaddam* a headache.' But the *muqaddam* replied, 'At least somebody knows how to speak around here.'

Hussein was frustrated because he saw the *muqaddam*'s visit as an opportunity to communicate with the Lebanese state, which, as the Arsalis maintained, had in the post-war years turned its back on the town. He and others believed that it was the role of their elected council to convey their needs and to 'correct' any idea that the quarrying industry had enriched Arsal. The current council had failed to mediate between the locals and the Lebanese state, leaving them at the mercy of immoral Syrian officials. Beyond their experience of corrupt local and state officials, my interlocutors imagined that an ideal incorrupt state existed – but one that was beyond their reach. Accessing this aspect of the state (Obeid 2010b) was deemed urgent at a time when residents found themselves facing increasing threats to their livelihoods. Their sovereignty was also at stake. Syrian practices in the town brazenly undermined residents' national pride, as the following narrative suggests.

By late January 2002, the build-up to the Iraq war was mounting. The war captured the attention and emotions of people in Arsal, as in other parts of the Arab world. They spent hours glued to the television, following Al-Jazeera's coverage of events in Iraq. Evening conversations were taken up with political analysis and the imagining of different political scenarios for Iraq and the region. One day, the secondary schools in the town organized a protest march. Students carried banners bearing slogans against the American invasion of Iraq. Some held aloft posters of Saddam Hussein, others of Palestinian leader Yaser Arafat. Pedestrians cheered them on from the side of the street, and the protest gradually swelled as people joined the march, which ended with a rally in front of the municipality building.

Just two days after the student march, a strange rumour began to circulate. It seemed to consolidate at once the pettiness and arbitrariness of Syrian rule. 'Keep this between you and me because the subject is very sensitive,' one man warned me. The Syrian intelligence officer, I was told, had called for an emergency council meeting. He apparently scolded the council members for not handing out posters of Hafez Al-Asad, the late Syrian President, to be carried in the student protest. The officer allegedly shouted at his audience: 'Had we raised dogs in this town, they would have been more loyal!' As far-fetched as this story sounded, it was repeated frequently over the next two days. My interlocutors were angered by this intervention. If the residents of Arsal were to carry posters celebrating 'Arab heroes,' in the current post-war climate the Syrian President would not be considered one of them, not here in Arsal anyway. Moreover, the idea of an officer rounding up the supposedly elite representatives of the town and calling them and their constituencies 'dogs' was degrading, especially if they were expected to show gratitude for the 'life of humiliation' (*dhull*) they were experiencing. It rubbed in the extent of Syrian domination.

If this incident could be dismissed as conspiracy theory based on hearsay, the announcement a few days later that the council was organizing the 'largest' protest against the war to be held in Arsal suggested otherwise. On my way to join the council-sponsored protest, I saw two local students who were studying in Iraq but had been called back home by their worried parents. They were to stay put, for a while at least, while the world held its breath and waited for the outbreak of war. They told me that they had refused to take part in the protest

FIGURE 13   Protest against the American invasion of Iraq

even though they felt furious about the imminent invasion. 'This protest,' one of them spat, 'is for Syria not Iraq!' They, too, had heard the rumours and were acting on them.

I found a spot where I could capture the arrival of the protestors on my camera. Leading the march were both the current and former mayor and some members of the council. A few rows behind them, massive posters of Al-Asad and large Syrian flags floated above the protesters. But the marchers also carried Lebanese flags and posters of other Arab leaders. The turnout was unprecedented, with thousands of people stretched out the entire length of the main road. While some people, like the two students, had resisted Syrian dominance by refusing to take part in the protest, others had decided to claim it back and voice their own views.

It is difficult to establish the accuracy of what did happen in the council meeting, or if, indeed, the Syrian officer had used the words now circulating in the town. What is clear, however, is that my interlocutors were vocalizing a particular sentiment in a political situation that was fast becoming untenable. By the end of 2003, the council was losing its credibility. The promising 'qualifications' of council members had proved deficient as they were accused of misusing *'ā'ila* to compromise the collective good to benefit their own interests. At national and regional levels, the council was seen to contribute to the marginalization of Arsalis as Lebanese citizens. People were left feeling betrayed by the actions of the council in conniving with the Syrians despite the growth of anti-Syrian sentiment in the town. It was soon realized that a council whose representation was based on *'ā'ila*, however fairly, was not sufficient or any longer appropriate to administer the town. But to what extent were people willing to relinquish *'ā'ila* as an idiom of political representation? The next section moves to the 2004 elections.

5 The 2004 Lists: 'Old Wine, New Bottles?'

The electoral lists drawn up in 2004 were not entirely dissimilar to those in previous elections. The Democratic Alliance (*Al-Tahāluf al-Dimuqrāti*) represented the Communist Party. Arsal's List (*Lā'ihat Arsal*), promoted by its instigators as the Lineage List, was organized by a former Jawhar MP. Arsal's Decision (*Qarār Arsal*) was constructed by the town's current Jawhar MP, under the aegis of the Syrian Ba'th Party. Finally, Justice and Development (*Al-'Adāla wa al-Inmā'*) represented a new, unprecedented group in the town, *Al-Jamā'a al-Islāmiyyah* (the Islamic Group). I will use 'Communist,' 'Lineage,' 'Ba'thist' and '*Jamā'a*' to refer to these lists, as the residents did.

The discourses that took place in the weeks preceding voting elaborate the 'trial and error' logic referred to by Ali, the political activist quoted at the start of this chapter. People sifted through previous failures and raised questions about the criteria that were necessary to equip the new council to administer Arsal in its current context. The NGO was an ideal place to hear the varying views of the list supporters since it offered typing and photocopying services specifically for the election, which brought the different political adversaries to our door. Across the board, the lists seemed to deploy common strategies. None of them included a candidate from the previous council. That would have been a clear disadvantage. Rather, each of the lists nominated only 'new people' who had not discredited themselves through dubious affiliations. Moreover, each of the lists ensured a fair representation of lineages, even the lists whose members preached against familism, such as the Communist List. From these two starting points, the advocates of the lists went on to promote their candidates' main qualities.

The Communists – their list name this time avoiding any association with cucumber – deployed their nationalistic history and invoked the memory of their heroism during the civil war, which was their principal currency. Zuhayr believed that the last council was an 'awakening' that would prompt voters to transcend their blind allegiance to their *'ā'ilat*. For this reason, he dismissed the Lineage List as a serious competitor. Like many others in the town, he also dismissed the Ba'thist List. 'No one would support the Syrians after the last council!' he insisted. 'What about the new contenders, the Islamists?' I asked. Basil, who was taking part in our discussion, contended that the Islamists had no place in Arsal. He was particularly angry with one of the local *shaykhs*, who had refused to conduct the Islamic prayers for the dead over the body of one of Basil's relatives, returned to Lebanon from Israel after Hizbullah struck a deal with the Israeli state for the exchange of prisoners and bodies. Apparently the sheikh decided that the Communist martyr was not a Muslim. This refusal seemed to infuriate many of my interlocutors, even those who were pious, who saw it as a dismissal of the martyr's heroism and an insult to Islamic virtues, for 'only God can determine who is a true Muslim.' The incident was utilized at the time of the election as a reminder of the 'narrow-mindedness' of the Islamists against the patriotism of the Communists and as a warning against the possible 'Islamization' of the town.

The Islamists, too, could take the nationalistic high ground. A few weeks before the election, their supporters also claimed a martyr, a 17-year-old who had smuggled himself abroad to fight in Baghdad and was returned as a martyr in 2003. This event, which brought glory to Arsal, was celebrated in a high profile rally as the body of the martyr was driven across the town. Campaigners for the Islamist List highlighted their candidates' pious and conscientious virtues, informed by their faith, a line of argument that resonated with many residents. Tired of the corruption that was impeding the progress of the town, many of my interlocutors hoped that the Islamists would bring a more ethical approach to governance. Abu Ahmad, who was not directly tied to the Islamic Group but seemed to support their list, argued that 'Candidates of *Jama'a* know God: they are *Allawī* [followers of Allah],' which, to him, implied honesty and integrity. They had also recently opened a new school, which Abu Ahmad saw as a much-needed step towards the town's development. Only two people on their list were official members of the Islamic Group. They had made this visible by donning the Islamist *abāya* and growing long untrimmed beards. The other candidates were 'ordinary' residents chosen for their 'decent reputation,' a quality that seemed essential to counter the corruption believed to pervade the previous council.

As for the Ba'thist List, rather than relying on the ideological arguments that other lists adopted, it deployed the unique strategy of promoting a clear developmental programme. It can be argued that this list was trying to appeal to voters who, by now, were unequivocally dismayed by the Syrians and their clients. They circulated leaflets that advanced a plan of action focusing on well-known priorities in the town, such as infrastructural work and the enhancement of public spaces. 'This is commendable,' Hana commented, as she inspected them. 'But if only the list was not Syrian! We want candidates who will do something for the town, we're not voting for the beauty of their eyes!' Cynically, her colleague responded that this list would win, 'programme or not,' given Syria's unquestionable power in the town and the region, and its previous record in manipulating elections. Hana was not completely convinced. 'True, they manipulate, but the elections are democratic. People are the ones who vote. What is scary is if they vote for their *'ā'ila* at the expense of the town.' If this were the case, then perhaps the Lineage List stood the best chance, as it did in the last election. But the power of *'ā'ila* on its own had proven insufficient, if not dangerous. And yet the lists had not yet renounced its centrality as an idiom of representation.

But why was *'ā'ila* still important in organizing the new lists when people were clearly voicing the need for new political avenues? This conundrum was best explained to me by the articulate political activist, Ali, mentioned above, using the metaphor of a horse: 'They all want to ride on the horse of *'ā'ila* because without it they cannot get anywhere.' In other words, lineage has become something like a common mode of transport, an independent tool for mobilizing constituencies regardless of ideological trajectories. For a list to stand a chance, it had better have a mixture of lineages, and this is precisely

LOCAL ELECTIONS: POLITICS AT THE MARGIN

FIGURE 14   Election leaflets for the Democratic Alliance

what all four lists offered. It was only then, having shown that they respected the traditional idiom, that the lists moved on to promote what they had to offer. 'But the horse on its own is not enough as it needs direction (*ittijāh*). The direction the lists will take the horse is what will make the difference.' Ali paused to think and continued, 'In the end, we all know who these lists are and who is behind them.' In other words, the various rhetorical devices utilized by the lists – 'nationalism', 'honesty', and 'development' among others – could not obscure their transparency. People knew what they represented. They were 'old wine in new bottles' and voters were not going to be duped by the new labels. Instead, Ali believed that what differentiated this election was that no list would win on its own. Rather, he anticipated that the council would be diverse because voters would elect 'those with *sira hasana*' (a decent reputation). Now that the integrity of political parties, traditional leaders, and previous council members had been put to the test, the process of trial and error had yielded a new criterion: people would look for candidates who were decent, defined as not corrupt, not known to have used state-provided authority and resources to build up patronage, and not having collaborated with the Syrians. This, he suspected, would drive voters to steer away from a bloc list and to vote for individuals who fitted the emerging criteria.

6 New Council, New Directions

The 2004 council was divided between an Islamic Group majority (13) and a minority from the Lineage List (8). All of the Communist and the Ba'thist candidates were rejected. Although Ali's prediction about the diversity of the council was not entirely accurate, his conception of the resilience of *'ā'ila* rang true. *'Ā'ila* maintained its position as the primary idiom of representation that acted as 'a means of transport' but not on its own (as exemplified by the loss of the Lineage List), and not without direction. Was the new direction one of political Islam? This question permeated post-election analysis. Several national newspapers linked Arsal's Sunni Muslim identity to the victory of the *Jamā'a*. But the residents' discourses about politics, as well as their complex strategies to alter their political realities, suggest an analysis that goes beyond merely their Sunni identity. Like other Muslims, my interlocutors had different emotional, intellectual, and political relationships to Islamism (Marsden 2007b; Eickelman and Piscatori 1996).

From the outset, the religious and political agendas of *Jamā'a* were downplayed both by its candidates and by the residents of Arsal. Their electoral list opted for a majority of citizens with no previous political affiliations.[11] Ali posited that perhaps the *Jamāʿa* were aware of Arsalis' critical sentiments towards radical Islam, given their leftist history; hence their focus on decency. 'Their candidates were persons who had a decent reputation in their neighbourhoods.' For example, one candidate, who was eventually voted mayor, was a respected school headmaster in his mid-thirties. Another, a man in his fifties, was an active member of the local Herders' Cooperative. These were examples of 'ordinary citizens' (*muwātin ʿādī*) who would work towards the good of their town with honesty and without the direction of a particular political party, even though they were on the *Jamāʿa* list. Robert Hefner ascribes the success of religious revivalist movements elsewhere to 'drawing themselves into mass society and away from exclusive elites' (2000: 219). In other areas of Lebanon, too, the *Jamāʿa* were believed to 'introduce their candidates on a non-sectarian basis, emphasising honesty and seriousness in their municipal work' (Hamzeh 2000: 744).[12]

In the context of Arsal, the *Jamāʿa* were new players starting with a clean slate. 'The council should be given a chance because if the *Jamāʿa* [candidates] are good people and their objective is the benefit of Arsal, then so be it,' Abu Abdu who was initially against this list concluded, possibly trying to reconcile himself to the election outcome. 'If they fail us,' he shrugged, 'we will not vote for them next time.' Perhaps he was elaborating the spirit of political experimentation that the residents came to adopt in local politics. Then again, his approach to the results reinforces the importance of understanding the entry of Islamist parties into the democratic political arena as that of any other player in electoral politics. An analysis that takes local history seriously reveals that their Islamism is not necessarily what gets them elected – in fact, one could argue, it was the de-emphasis of Islamism that allowed for their election.

Abu Abdu's comment confirms a common belief that elections have the power to create transformation. My interlocutors differed on the interpretation and desirability of this transformation. For instance, Zuhayr and his secular counterparts considered the results a 'shame' (*ʿār*) that tinged Arsal's leftist history, a rupture that pulled the town backward. Others, however, suggested a

---

11   I knew of members who had previously had different affiliations, but none were on the 2004 list. For example, I knew one man who was a Communist till the late 1990s but joined the *Jamāʿa* afterwards. His friends accused him of doing it for money, especially that he had adopted Islamic clothing and no longer shook hands with women.

12   Other Islamist groups in Lebanon, such as Hizbullah, are known to use democratic discourses as part of their electoral strategy to work within the 'rules of the Lebanese political game' (Harik 2004: 109, see also Norton 2007 and Deeb 2006).

different reading that related to the town's location at the border of the nation-state, both literally and figuratively. The most immediate outcome of the elections was that Arsalis were able to defeat the Syrians through democracy, by ensuring they voted out the Syrian-sponsored list. The residents were performing a 'politics of presence' (Brink-Danan 2009) through elections that had the power to induce change. The victory was only a partial one as the Syrian apparatuses maintained their effective grasp on local affairs and only withdrew from Lebanon a year later as a result of national and international pressures in the aftermath of the assassination of Prime Minister Hariri.

The victory of *Jamā'a*, I suggest, is best understood in terms of their emergence as an alternative that blended local forms of political representation (through the upholding of *'ā'ila*) with prospects of accountable governance, both crystallized in idioms of decency that sought the moralization of politics. This contention helps us look beyond the 'Islamization of politics,' a conclusion that links the results to the residents' sectarian identity but overlooks the intersections between the town's history with the state, its border predicament in the post-war context, and the residents' local political idioms. In the post-war years, Arsalis tolerated the ambivalent predicament of being governed simultaneously by the apparatuses of two states, the Lebanese and the Syrian, whose fluctuating political relations had a bearing on everyday life. The 'structural effects' (Mitchell 1991) of the mundane practices of the Syrian state threatened the sovereignty of the residents. The elections served as a site for Arsalis to establish themselves as Lebanese citizens and to deploy a democratic 'practical form' (Paley 2002: 471) to induce changes within the existing power structures. An understanding of these complex connections points to a dynamic attempt on the part of the electorate to fulfil the original promise of democracy by debating and negotiating the process of representation to bring about change.

Um Abdu was pleased. The candidate she had voted for won; he was an honest, decent neighbour, she commented, and that's what the town needed. The Syrians were 'given a message: they were not welcome any more.' But she was more pleased that the Arsalis had given this message peacefully and democratically (*bil dimuqratiyya*). She reminded me that 'today is different from the past' in that people longed for *'aqlāniyya* (rationality), for humans moved forward, not backward, she maintained. That democratic processes honoured the local value of kinship in politics as they left room for people to protest unjust and unwanted forces was for her and others evidence of the triumph of morality.

CHAPTER 8

What the Future Hides

Hana was preoccupied. Her supply of fabric dye had run out and she needed to find more if her team at the carpet workshop was to meet the deadline for a substantial order of *kilim* rugs. The border had been closed for weeks, ruling out her usual trips to Syria, where she had established long-term relationships with merchants. While individual travellers were still allowed to enter Syria, they could not bring goods in or out of the country for security reasons. Like many others who relied on cross-border visits to run their businesses and to shop economically, Hana worried that the cost of finding alternative suppliers would leave her workshop out of pocket. She explained her concerns as she prepared a glass of *mate*. 'It's not that you can't find this stuff in Lebanon, but it is expensive,' she said. 'We will hardly make a profit paying [Lebanese] prices.' 'I'd rather we all died of starvation than rely on Syria ever again!' her brother Yusif retorted, continuing defiantly, 'Let them shut the border. Things will change now.' His cousin Imad, who was grabbing the *mate* cup from Hana, challenged him sarcastically: 'Ha! Things will change? Why? Because [Saad] Hariri is coming?'

The assassination of Prime Minister Rafik Hariri in 2005 had been responsible for the closure of the border by the Syrian authorities. For about three months they suspended the export of Lebanese agricultural produce, and as the only other exit route was through Israel, which was not an option, long queues of internationally licensed trucks carrying perishable produce had been stranded at the border crossings. Syrian security checks also multiplied in the highlands, substantially curbing smuggling operations. These punitive measures were taken in retaliation for accusations that the Syrian state had masterminded the assassination, a claim made by the so-called Lebanese March 14 Alliance.[1] Crowds took to the streets of Beirut, prompting the opposition in parliament to push for the implementation of UN Resolution 1559, issued in 2004, calling for the evacuation of foreign powers (i.e. Syria) from Lebanon and for the disarming of militias (i.e. Hizbullah).[2] The protests succeeded, where no previous

---

1 The Alliance is named after the date of what became known as *Thawrat al-Arz* (the Cedar Revolution).
2 For a full text of the resolution, see the following UN press release: http://www.un.org/News/Press/docs/2004/sc8181.doc.htm.

initiatives had, in overthrowing the pro-Syrian government. Crucially, they resulted in the withdrawal of Syrian troops from Lebanon.

A sense of vindication permeated Arsal, as it did all anti-Syrian groups in Lebanon. Were the residents going to be treated as Lebanese citizens with rights at last? Was there hope for the Lebanese state? But in the months that followed there developed a long-lasting polarization of government and politics into two opposing camps: an anti-Syrian camp led by Hariri's son Saad, and a pro-Syrian camp dominated by Hizbullah. During the post-war years Yusif, as a former Communist, had shunned Hariri's politics and what they stood for: 'Capitalism (*ra'smāliyya*) and social division (*tafriqa ijtimā'iyya*).' But in the post-Syrian context, he and others in the town welcomed the prospect of support from Saad Hariri's office in Beirut. For the first time, Hariri's recently established Future Movement Party (*Al-Mustaqbal*) began to operate in Arsal and its representatives spread the word that Saad was going to reward Arsal for its perseverance under the oppressive Syrian rule, the same rule that was believed to have killed his father. The town might finally have a national patron, a Sunni leader, to be sure, who would advance its needs and make its voices heard from his vantage point at the centre of politics. Hariri, Yusif told us, was going to invest in Arsal. This rumour was not to be taken lightly, given the Hariri family's famous affluence and influence within and outside the country. All it took, Yusif suggested, was a project (*mashrū'*) on a large enough scale to accommodate an Arsali labour force and open up its market: a dairy factory, perhaps, or a food factory that would capitalize on Arsal's fruit production. 'Aren't they called *Al-Mustaqbal*?' Yusif wondered. 'Let them bring us a new *mustaqbal* (future) so our lives can be normal again.' Sensing Yusif's slight lack of conviction, Imad interrupted: 'Why don't you leave the future for now? Only God knows what the future hides. Let's stay in the present, please.'

Stef Jansen (2015) invites us to 'pry open' what 'normal' means to our interlocutors, who, when experiencing different kinds of rupture, tend to express a 'yearning for normal lives.' The end of the 1975–1990 war had brought hope that life would return to some sort of 'normal' after years of violence and disruption, and after an absence of formal channels of governance that had left the town, and Lebanon, in a dire state. In the early 1990s, 'a better life' (*'isha afdal*) was a generic aspiration articulated by my Arsali interlocutors, in common with other Lebanese. This was seen to be the remit of a functional state that would, after gradually rebuilding its institutions, 'return' to providing necessary services for its citizens, thus granting them a dignified life. Regulated livelihoods, schools for their children, health provisions for old and young, roads that would connect them to the rest of the country, and jobs that would occupy their young men; all the services that would re-centre their border lives. But a normality framed in these terms never really transpired. This was commonly ascribed to the 'weakness' attributed to an ailing Lebanese state, incapable of recovery from a damaging war and unable to exercise its sovereignty, given its post-war subservience to the Syrian state. Where it had clout, its attention was divided, favouring the capital over the regions and operating through an entrenched system of clientelism. In their yearning for 'normal,' my interlocutors expressed a desire for an ideal state and hoped to be incorporated into its field of vision: 'If only the state would *tittallaʿ fīna* (look toward us)!' (Obeid 2010b).

Arsal's location on the ambivalent northeastern border contributed to the residents' marginality, reinforced by different historical episodes. The abuses practised by Syrian officials in the few years before the Syrian withdrawal reconfirmed their 'unequal location as well as unequal relations' (Green 2005: 1). While there was recognition that the entire country had been brought to its knees by Syria's dominance, the Arsalis claimed exceptionalism in their suffering. 'No one felt the weight of Syria like us Arsalis!' my interlocutors would repeat, emphasizing the 'remote' and 'forgotten' predicament of their border lives. But the withdrawal of the Syrian state from Lebanon changed the mood of the country as people contemplated their future with hope. Did a better *nasīb* (fate), to use the local term, await them now? Yusif's conversation with his cousin suggested a link between the unknown, the *mustaqbal*, and a particular political leadership that might lend some predictability to the future, one that would reconstitute 'normal lives' by mitigating their border location and reconnecting them to 'the heart of things.' In response to his cousin's warning that the future was not legible, Yusif narrowed down his own reading to two possibilities: 'Leave God aside. What *could* the future hide? Either we will move forward, or we will move backward.'

The euphoria following the Syrian withdrawal was coupled with concern over the future that awaited the town, the Biqaʿ, and the country. Conversations speculating on all the possibilities that lay ahead, whether in the near or distant future, animated the town. My interlocutors felt they were entering a new phase, a transitional one (*marhala intiqāliyya*), from 'the 'Syrian time' (*waqt al-Suriyyīn*) – the post-war years up until 2005, now a thing of the past – to a new time. How this time would depart from the past and present remained to be seen.

1   A Visit in Post-Syrian Time

I visited Yusif's house about a year after this conversation. When I asked him how things were, his response was full of resignation. *'Ala – hattit īdik*,' meaning

'Exactly where you left us', an expression I was to hear over and over again during that visit as people communicated their uneasy sense of stagnation. They had moved neither forward nor backward, they were saying. The awaited *mashrū'* (project) hadn't materialized, nor was it on the horizon. Hariri, I was told, had invested in the end, but only by making a nominal contribution to a small-scale project. Yusif thought this was no more than a gesture to say that Arsal was on his radar. 'They are all the same,' he sighed. Political leaders were only in it to advance their own agendas. His analysis suggested that 'adopting' Arsal chimed with the March 14 Alliance's call for the Lebanese state, through the army, to control the border with Syria. The call became more urgent in the months after the Prime Minister's death as a series of assassinations and explosions shook the capital and brought back memories and fears of the civil war (Hermez 2017).[3]

Although the residents themselves had supported this call to securitize the border only a year before, I learnt that the quarries were working again, that people were crossing the border to Syria to run errands and to visit Syrian markets, though not with the same frequency, and that smuggling had resumed. '*Ādī* (it's normal)!' Jamila commented as we caught up over tea and spinach pies at the NGO. 'It's like we were before, but without the Syrians.' This return to the not-so-normal 'normal' of 'Syrian time' points to the disconnection between the political elite in the capital and people living on the margins of the nation-state. It also brings to light the temporality of the border and how quickly people's practices and attitudes around it can transform, or not. While a clean separation from Syria – territorially, politically, and economically – was an ideological imperative, it was not easily applicable given the porous nature of the border and the long-standing social, cultural, and economic ties that connected the border communities (both in the northwest and the northeast of Lebanon) to Syria. Yusif felt slightly betrayed that his town was being used for political purposes. Hariri had played a cheap card to rally support in the town, Yusif believed. The 'sect card' (*waraqat al-tā'ifa*) was bound to draw people in, given how political factionalism in Lebanon was intertwined with sectarian affiliations. Having Hariri as their Sunni *za'īm* (leader) was a prospect that promised protection and prosperity to the marginalized Sunnis of the

---

3 In December 2005, following explosions and killings in Lebanon that year, the Lebanese government requested the creation of a UN tribunal of 'international character' with a mandate to try those responsible for the assassination of Hariri and to further investigate the killings that followed. This led to the establishment of the Special Tribunal for Lebanon in January 2007. For details, see https://www.stl-tsl.org/en/about-the-stl/636-creation-of-the-stl.

Biqaʻ. 'But what has he *actually* brought to the town?' Yusif wondered. Already, the future was resembling the past.

Others sustained their hopes. Things took time, they argued. Hariri was a decent leader, he was new to the game, and all he needed was time (*waqt*) as well as people's faith. After all, he was still facing the Syrians, some reasoned. The troops had departed but support for the Syrian state still dominated the government (and perhaps the country). Certainly this was the position in the Northern Biqaʻ, where Hizbullah was the dominant political power. 'We have to be patient,' Abu Hasan told me. 'But Arsal may pay a price. We are alone in the meantime.' He was referring to the fact that the town now stood out because of its political/sectarian position in relation to its pro-Hizbullah and pro-Syrian neighbours. His daughter Khadija elaborated on this point. During the Cedar Revolution, she recounted, Arsal's conspicuous involvement in the protests in Beirut provoked their pro-Hizbullah neighbours.

> Every bus, van and car filled up with Arsalis and drove down to Beirut to take part in the protests. Some people had never even set foot in Beirut. But everyone wanted the Syrians out! The young, the elderly, everyone went. The town was so quiet; you'd think the birds went to the protests too!

This sentiment, however, was not shared in the northern Biqaʻ. Khadija described how crowds of angry people in the valley flanked the main road and began to throw stones at Arsali vehicles returning from the protests. 'The Shiʻas wanted to kill us! We are stuck here in this difficult situation.' To drive home her point that the divisions had become irreconcilable, she continued, 'They are here [pointing in one direction] and we are there [pointing in the opposite direction]. Our minds will never meet.'

These inter-regional tensions were now explicitly expressed in political sectarian language that constructed an Arsali Sunni community against a Shiʻi one.[4] This language soared during and after the 2006 July war in which Israeli Defence Forces (IDF) launched an attack on Lebanon for 34 days in retaliation for one of Hizbullah's cross-border military operations. The IDF targeted the south of Lebanon and the southern suburbs of Beirut, destroying entire villages in the south and neighbourhoods in the capital. Airstrikes targeted the

---

4 Ethnographers of Lebanon have shown that sectarianism, even though it appears to be primordial and is talked about as such in the media and public circles, is contingent on a variety of institutions and services that reproduce and recalibrate experiences of sectarian belonging (Joseph 1999; Cammett 2014; Harb 2010; Nucho 2016). In a similar vein, post-war expressions of sectarianism in Arsal cannot be understood outside of post-war politics and the way they played out in particular ways in the Northern Biqaʻ.

airport and the Beirut-Damascus highway to block access to Syria. The Biqaʿ region was also hit, as IDF planes shelled the highlands to prevent the possibility of smuggling weapons from Syria. The war entrenched political and sectarian divisions in the country, particularly when the government initially held back from the war, blaming Hizbullah for provoking it, before taking action nationally and internationally as the disproportionate effects of the Israeli attacks became visible.[5]

When I paid the town a visit two months after the war, my interlocutors complained that relations with their neighbours had worsened. Abdu angrily recounted how, during the July war, their neighbouring villagers planted a sign pointing to the town that read *Israʾīl min huna* (This Way to Israel), a cause of grave insult to the residents who still took pride in their town's reputation as *Umm Al Shuhadāʾ* (Mother of Martyrs), earned during the 1982 Israeli invasion. Abdu explained that their neighbours were angered because Arsali residents continued to smuggle diesel from Syria. They were seen to be opportunists, taking advantage of the war to make a profit. But Abdu claimed that while Arsali smugglers 'did carry on with their work,' they were also engaging in acts of heroism. 'We are the only ones who supplied hospitals with diesel [for their generators]. We did that under the shelling! They should be thanking us rather than calling us traitors (*yu-khawwinūna*)!' On a later visit that winter I heard more stories that revealed how political tensions at the national level were inflected in the region. One afternoon, I heard Siham reprimand her ten-year-old daughter for offending a classmate in the Shiʿi school that she and her siblings attended in the valley. Siham was summoned by the school principal to inform her that her daughter had called Sayyid Hasan Nasrallah, Hizbullah's Secretary General, '*Sayyiʾ* Hasan.' Sayyid is the honorific title that precedes the leader's name, denoting descent from Prophet Muhammad's lineage. *Sayyiʾ*, however, means 'bad' – 'bad Hasan'. Her daughter defended herself by saying that 'She started it! She called Saad [Hariri] ugly!' Later Siham and I laughed at this story. 'Children!' Siham remarked, shaking her head at their ability to pick up what was said and felt around them. 'But seriously, this is worrying. These are scary times. If children are fighting over Nasrallah and Hariri, what can we expect from the country? This has become normal for us, they (Nasrallah and Hariri) fight there and we fight here.'

Siham's comment is an apt portrayal of the connectivity of the local with the national. The polarization of politics in the aftermath of Hariri's assassination,

---

5   Hizbullah later accused the government and the March 14 Alliance of encouraging this war in a bid to disarm and weaken Hizbullah. See Paul Salem's (2008) analysis in this report: http://carnegieendowment.org/files/PaulSalemChapter.pdf.

intensified further by the 2006 July war, manifested itself in complex ways across the country. In the 'post-Syrian time' my Arsali interlocutors felt more isolated in the political landscape of the Northern Biqaʿ region as political differences played out in particularly sectarian tones. While I heard stories of fights, disagreements, and uncomfortable interactions and snubs between Arsalis and their neighbours, many others felt (or hoped) that this was a fleeting moment, a necessary phase even, after the upheavals that the country had gone through in less than two years. Abu Ali, a seasoned resident who had participated in the 1958 war in which Arsali fighters killed some of their Shiʿi neighbours, insisted that in the end, the social bonds accrued over time (*ʿishra*) would prevail between Arsal and its neighbouring villages,[6] whether in Lebanon or Syria. Abu Ali garnered authority from his long life that had witnessed two civil wars, a number of Israeli invasions, and the arrival and withdrawal of Syrian forces, among other ruptures that shaped and reshaped the landscape of his town and country. From his wide purview, the frictions of the moment concealed the fluctuations of human relationships and made them appear eternal or primordial. 'The prophet has asked us to look after the seventh neighbour,' he reminded me, adding that the prophet never specified the sect or denomination of that neighbour. In the larger scheme of things, he was suggesting, these tensions were a storm that could only end in calm.

The rising political factionalism in the country, propagated by the polarization of politics into two camps (anti-Syrian and anti-Hizbullah versus pro-Syrian and pro-Hizbullah), intertwined with sectarian identification in the aftermath of the Syrian withdrawal, thus point to the dynamic and processual nature of sectarianism. In Arsal, the withdrawal of Syrian troops brought hope for a different reality, that same yearning for a peaceful life in which a functional state would provide its citizens with 'normal lives.' Joanne Nucho shows how, in Lebanon, the relationship between infrastructures and services, the very things that make life 'normal,' 'are produced by sectarian, political and religious organisations at the same time that they are the channels through which sectarian belonging and exclusion are experienced, produced, and recalibrated' (2016: 6). From the marginal perspective of the border, a connection

---

6 In an interview I had recorded with Abu Ali in the mid-1990s, he made a similar point saying that the ethos of neighbourliness in the 1958 war served as a buffer against the violence between villages. 'When we won that battle against [villages in] the valley, we confiscated their weapons and killed about five people. We could have killed fifty but the Arsalis knew that neighbours would always be neighbours. They give us tomatoes; we give them wheat. So we knew we should stay on good terms with them. Also, there is *qarāba nisāʾiyya* (female kinship, marriage ties) and friendships between Arsal and these villagers. This played a great role in limiting the deaths so that we can maintain neighbourly ties.'

with central power through a national Sunni *zaʿīm* in the figure of Hariri seemed like the best way to 'turn the attention' of 'the state' towards the town. But this projection was too hasty. Even when Saad Hariri was eventually elected as Prime Minister in 2009, the town remained marginal to his politics. In the turbulent meantime, my interlocutors carried on with their border lives and, like the rest of the country, pondered about what the future hid for them.

2   Is it Possible to Move Backwards?

It is April 2015. I have returned to Arsal after an absence of three years. The town has been in the spotlight for months. The Syrian protests that broke out in the upshot of the 'Arab Spring' had turned violent and news, analysis, and reports were preoccupied with how the conflict in Syria was 'spilling over' to Lebanon. The effects of the war were felt across the country. More than 1.3 million refugees from Syria had fled to Lebanon. The UNHCR estimated that this was equivalent to more than a quarter of Lebanon's resident population.[7] The government was struggling to control the increasing influx of displaced people fleeing the war across the border. The overwhelming solidarity with Syrians that Arsali residents showed at the beginning of the protests soon began to wane. In less than four days in November 2013, 12,000 refugees crossed the border to the town during the Battle of Qusayr. Within less than two years, the number of refugees had risen to an estimated 90,000,[8] more than trebling the population of Arsal, which, as I mentioned in the Introduction, was estimated at 32,000 in 2003. The highlands became a battleground between different Syrian factions, and by 2014 Islamic State (Daʿish) and Al-Nusra Front had established operational bases on people's farmlands and orchards. The highlands became a dangerous place, subjected to regular bombardment by the Syrian regime. News of abductions, murders, and fighting in the highlands began to threaten everyday life. Although Hizbullah fighters had been operative in Syria since 2012, the Party openly announced its support for the Syrian Armed Forces (and state) and fought rebel groups on Syrian land in Qusayr. This created a remarkable political controversy in Lebanon and fuelled the divisions that had been growing since Prime Minister Hariri's assassination.

---

7   See http://www.unhcr.org/pages/49e486676.html#.
8   Locals working on the ground with refugees informed me that the UNHCR officially registered 42,000 refugees. But they claimed that the number was much higher, reaching 90,000 since many who came from Syria were living in houses and basements rather than the newly built camps. They argued that these people were not counted.

Arsali officials appeared on television, calling for the intervention of 'the state' and for the army to control the border. In the meantime, the border remained open, 'indescribable' (Reeves 2014: 3) in its lack of territorial division between the two countries, until August 2014.

An attack on the town by Da'ish during that month prompted new measures of 'bordering' (Green 2012). For four days the town was under siege as fighters ransacked a police station, attacked army barracks, kidnapped Lebanese security personnel, and spread terror in the town. In an extreme and urgent measure, the Lebanese army with the collaboration of Hizbullah shelled the fighters who were driven back into the highlands. The price of this victory was the destruction of local homes, the burning down of an entire Syrian refugee camp (with a death toll of 50 people), and the loss of about 20 Lebanese soldiers. It was only then that the army was deployed in new security measures that included closing off the edge of the town from the highlands with barracks and checkpoints. The entrance of the town was also securitized with two new successive checkpoints, the first controlled by the army, the other by the Lebanese *mukhābarāt* (security services). The only two town exits were now enclosed.[9] These new security measures aimed at targeting the Syrian refugee population. No Syrian national was allowed to enter or leave the town without registering at both mentioned checkpoints. Locals were subjected to these same checks. Foreigners and visitors without official papers were denied entry.

I visit Yusif's house and it's an emotional reunion. My last visit seemed like another era. How were they coping without access to the highlands? How did they feel about the new security measures? What had happened to the quarries? What about their orchards? 'It's all gone,' Yusif answers.[10] I listen sympathetically as Yusif and his family describe the changes in the town: the physical transformation of the neighbourhoods as they filled up with different sized refugee camps; the unprecedented volume of people on the streets; the tensions between the locals and the Syrian refugees; the loss of livelihoods with the abandonment of orchards and quarries; the loss of mobility across a previously open border and within Lebanon; anxiety over what fate awaits the town and its people. The repercussions of the Syrian protests, I am told, left Arsal abandoned. National and social media depicted the town as a haven for radical Islamists. If there were sympathetic voices, they did not garner enough

---

9   This kind of security arrangement is not unheard of in Lebanon. In fact, it very much resembles the way Palestinian camps, considered to be impenetrable 'security islands' (Ramadan 2009: 158), have been handled across the country.

10 I learnt that lands up to the checkpoint could still be accessed. So some quarries were still operating. But beyond that, security officers advised residents not to go, or to go at their own risk.

political will to find solutions for the locals, who felt they bore the brunt of the 'spillage' from the Syrian crisis. What were the idle *shabāb* (young men) going to do without their land and without jobs? If the Syrian refugees were not to return, how would Arsal's schools, infrastructures, and services accommodate the increase in population? 'Only God knows what the future hides,' Yusif sighs. I remind him playfully of his projections in previous years. 'There are only two possibilities. Either you will move forward or you will move backward,' I say. 'A-hoo! We have definitely moved backwards, by about one hundred years,' he responds. 'We haven't just moved back to the past, we have been catapulted (*naqafūna*) out of history altogether.'

The long-term effects of these transformations on the town and its sociality are beyond the scope of this analysis. The ramifications of what is now known as '*al-azma al Sūriya*' (the Syrian crisis) are still unfolding as I finish this book. But the questions contemplated by Yusif and other interlocutors speak to some of the questions explored in the preceding chapters, for they are ultimately about 'changing times' that demand 'figuring out,' so that residents could carry on (*yu-maddū*) with their lives. The book has shown that these lives have always been entangled with the border location of the town and the regional and national political landscapes that shape them. Yusif's comment about being 'catapulted out of history' may be his way of expressing an unprecedented predicament, unfamiliar in its scale to anything he or the older generation in his town had experienced. This is worse, he was saying, than backwardness itself, imagined as 'going back in time.' For as he and others believed, 'humans should move forward not backward.' The images of the town as a place that needed to be controlled, tamed, and warded off from the rest of the country reinforced the trope of the 'frontier,' 'savage, primitive and unregulated' (Volk 2009: 264), that had coloured the history of the region in general and Arsal in particular. The enforced lack of mobility, the closures that replaced the town's openness to the world in the earlier post-war years, the safety and security that nourished 'opened doors,' 'togetherness,' and mutual engagement were the very features of a progress that now seemed threatened. If the future is meant to be better than the present and the past, then the linearity of history was in question. Then again, life as it stands might become 'normal' as people *yu-maddū* with their border lives.

Bibliography

Abrams, P. (1988). 'Notes on the Difficulty of Studying the State.' *Journal of Historical Sociology* 1(1): 58–89.

Abu-Lughod, L. (1986). *Veiled Sentiments. Honour and Poetry in a Bedouin Society.* Berkeley, Los Angeles and London: University of California Press.

Abu-Lughod, L. (2008). *Dramas of Nationhood: The Politics of Television in Egypt.* Chicago: University of Chicago Press.

Abu-Lughod, L. (2009). 'Dialects of Women's Empowerment: The International Circuitry of the Arab Human Development Report 2005.' *International Journal of Middle East Studies* 41(1): 83–103.

Abu Rabia, A. (1994). *The Negev Bedouin and Livestock Rearing: Social, Economic and Political Aspects.* Oxford: Berg Publishers.

Ahearn, L. (2001). *Invitation to Love: Literacy, Love Letters and Social Change in Nepal.* Michigan: The University of Michigan Press.

Alatas, S.F. (2006). 'A Khaldunian Exemplar for a Historical Sociology for the South.' *Current Sociology* 54(3): 397–411.

Al-Jamr, A. (2003). *Assafir Newspaper.* July 15.

Allerton, C. (2007). 'What Does it Mean to Be Alone?' In R. Astuti, J. Parry and C. Stafford (eds.). *Questions of Anthropology.* Oxford and New York: Berg: 1–28.

Amit, V., S. Anderson, V. Caputo, J. Postill, D. Reed-Danahay and G. Vargas-Cetina (2015). 'Thinking through Sociality: The Importance of Mid-Leven Concepts An Anthropological Interrogation of Key Concepts.' In V. Amit (ed.). *Thinking Through Sociality: An Anthropological Interrogation of Key Concepts.* Berghahn Books: 1–20.

Anderson, S. (2015). 'Sociability: the Art of Form.' In V. Amit (ed.). *Thinking Through Sociality: An Anthropological Interrogation of Key Concepts.* Berghahn Books: 97–127.

Aretxaga, B. (2003). 'Maddening States.' *Annual Review of Anthropology* 32(1): 393–410.

Baalbaki, A. (1997). 'Transformations in the Pastoral Nomad System in the Village of Arsal.' *Periodicals of the Institute of Social Sciences.* Lebanese University, Beirut 4: 67–84.

Barbour, B. and P. Salameh (2009). 'Consanguinity in Lebanon: Prevalence, Distribution and Determinants.' *Journal of Biosocial Science* 41(4): 505–517.

Barlocco, F. (2010). 'The Village as a "Community of Practice": Construction of Village Belonging Through Leisure Sociality.' *Bijdragen tot de Taal-, Land- en Volkenkunde* 166(4): 404–425.

Baroudi, S.E. (2005). 'Lebanon's Foreign Trade Relations in the Postwar Era: Scenarios for Integration (1990–Present).' *Middle Eastern Studies* 41(2): 201–225.

Baumann, H. (2017). *Citizen Hariri: Lebanon's Neoliberal Reconstruction.* Oxford, New York: Oxford University Press.

Baxter, D. (2007). 'Honor Thy Sister: Selfhood, Gender, and Agency in Palestinian Culture.' *Anthropological Quarterly* 80(3): 737–775.

Bell, S. and S. Coleman (eds.) (1999). *The Anthropology of Friendship*. Oxford, New York: Berg.

Berliner D., M. Lambek, R. Shweder, R. Irvine, and A. Piette (2016). 'Anthropology and the Study of Contradictions.' *HAU: Journal of Ethnographic Theory* 6(1): 1–27.

Beydoun, A. (1992). 'The South Lebanon Border Zone: A Local Perspective.' *Journal of Palestine Studies* 21(3): 35–53.

Bocco, R. (2000). 'International Organizations and the Settlement of Nomads in the Arab Middle East, 1950–1990.' In M. Mundy and B. Musallam (eds.). *The Transformation of Nomadic Society in the Arab East*. New York: Cambridge University Press: 302–334.

Booth, W.J. (1993). 'A Note on the Idea of the Moral Economy.' *The American Political Science Review* 87(4): 949–954.

Borbieva, N.O. (2012). 'Kidnapping Women: Discourses of Emotion and Social Change in the Kyrgyz republic.' *Anthropological Quarterly* 85(1): 141–169.

Bourdieu, P. (1966). 'The Sentiment of Honour in Kabyle Society.' In J.G. Peristiany (ed.). *Honour and Shame: The Values of Mediterranean Society*. Chicago: University of Chicago Press: 191–242.

Bourdieu, P. (1977). *Outline of a Theory of Practice* (Vol. 16). Cambridge: Cambridge University Press.

Brandell, I. (2006). 'Introduction.' In I. Brandell (ed.). *State Frontiers. Borders and Boundaries in the Middle East*. London, New York: I.B. Tauris: 1–32.

Brink-Danan, M. (2009). '"I Vote, Therefore I Am." Rituals of Democracy and the Turkish Chief Rabbi.' *PoLAR: Political and Legal Anthropology Review* 32(1): 5–27.

Butt, B. (2011). 'Coping with Uncertainty and Variability: The Influence of Protected Areas on Pastoral Herding Strategies in East Africa.' *Human Ecology* 39(3): 289–307.

Cammett, M. (2014). *Compassionate Communalism: Welfare and Sectarianism in Lebanon*. Ithaca and London: Cornell University Press.

Carrithers, M. (2008). 'From Inchoate Pronouns to Proper Nouns: a Theory Fragment with 9/11, Gertrude Stein, and an East German Ethnography.' *History and Anthropology* 19: 161–186.

Chatty, D. (1980). 'Changing Sex Roles in Bedouin Society in Syria and Lebanon.' In N. Keddie, L. Beck (eds.). *Women in the Muslim World*. Cambridge: Harvard University Press: 399–415.

Chatty, D. (ed.) (2006). *Nomadic Societies in the Middle East and North Africa: Entering the 21st Century* (Vol. 81). Leiden: Brill.

Clarke, M. (2009). *Islam and New Kinship: Reproductive Technology and the Shariah in Lebanon*. Oxford: Berghahn Books.

Coles, K. (2004). 'Election Day: The Construction of Democracy through Technique.' *Cultural Anthropology* 19(4): 551–580.

BIBLIOGRAPHY

Collier, J.F. (1997). *From Duty to Desire. Remaking Families in a Spanish Village*. New Jersey: Princeton University Press.

Collier, J.F. and S.J. Yanagisako (1987). 'Introduction.' In J.F. Collier and S.J. Yanagisako (eds.). *Gender and Kinship. Essays Toward a Unified Analysis*. Stanford California: Stanford University Press: 1–13.

Coppolillo, P.B. (2000). 'The Landscape Ecology of Pastoral Herding: Spatial Analysis of Land Use and Livestock Production in East Africa.' *Human Ecology* 28(4): 527–560.

Corsín Jiménez, A. (2003). 'On Space as a Capacity.' *Journal of the Royal Anthropological Institute* 9(1): 137–153.

Creed, G.W. (2000). '"Family Values" and Domestic Economies.' *Annual Review of Anthropology* 29: 329–335.

Darwish, M.R., S. Hamadeh and M. Sharara (2001). 'Economic Sustainability of Dry Land Use: The Case Study of Irsal, Lebanon.' *Journal of Sustainable Agriculture* 17(4): 91–102.

Darwish, T., C. Khater, I. Jomaa, R. Stehouwer, A. Shaban and M. Hamzé (2011). 'Environmental Impact of Quarries on Natural Resources in Lebanon.' *Land Degradation & Development* 22(3): 345–358.

Das, V. and D. Poole (2004). 'The State and its Margins.' In V. Das and D. Poole (eds.). *Anthropology in the Margins of the State*. Santa Fe, NM: School of American Research Press: 3–33.

Deeb, L. (2006). *An Enchanted Modern: Gender and Public Piety in Shi'i Lebanon*. Princeton: Princeton University Press.

Deeb, L. and J. Winegar (2012). 'Anthropologies of Arab-majority Societies.' *Annual Review of Anthropology* 41: 537–558.

Deek, C. (2003). 'Adaptation Strategies of Small Ruminants Production Systems to Environmental Constraints of Semi-arid Areas of Lebanon.' *Unpublished Thesis*, American University of Beirut, Lebanon.

De Grazia, V., and E. Furlough (eds.) (1996). *The Sex of Things: Gender and Consumption in Historical Perspective*. Berkeley: University of California Press.

Delaney, C. (1995). 'Father State, Motherland, and the Birth of Modern Turkey.' In Sylvia Yanagisako and Carol Delaney (eds.). *Naturalizing Power. Essays in Feminist Cultural Analysis*. New York: Routledge: 177–200.

Dent, C.M. (2001). 'ASEM and the "Cinderella Complex" of EU-East Asia Economic Relations.' *Pacific Affairs* 74(1): 25–52.

Duben, A. and C. Behar (1991). *Istanbul Households: Marriage, Family, and Fertility 1880–1940*. Cambridge: Cambridge University Press.

Edwards, J. and M. Strathern (2000). 'Including Our Own.' In Janet Carsten (ed.). *Cultures of Relatedness: New Approaches to the Study of Kinship*. Cambridge: Cambridge University Press: 149–66.

Eickelman, D. and J. Piscatori (1996). *Muslim Politics*. Princeton: Princeton University Press.

El-Khazen, F. (2003). 'Political Parties in Postwar Lebanon: Parties in Search of Partisans.' *Middle East Journal* 57(4): 605–624.

El Nour S., C. Gharios, M. Mundy and R. Zurayk (2015). 'The Right to the Village? Concept and History in a Village of South Lebanon.' *Justice Spatiale | Spatial Justice* 7: 1–24.

El-Solh, R. (2004). *Lebanon and Arabism: National Identity and State Formation*. London, New York: I.B. Tauris.

Elliot, A. (2016). 'Paused Subjects: Waiting for Migration in North Africa.' *Time & Society* 25(1): 102–116.

Ellis, F. (2000a). *Rural Livelihoods and Diversity in Developing Countries*. Oxford: Oxford University Press.

Ellis, F. (2000b). 'The Determinants of Rural Livelihood Diversification in Developing Countries.' *Journal of Agricultural Economics* 51(2): 289–302.

Forte, T. (2001). 'Shopping in Jenin: Women, Homes and Political Persons in the Galilee.' *City & Society* 13(2): 211–243.

Galaty, J.G. and P. Bonte (1991). *Herders, Warriors, and Traders: Pastoralism in Africa*. Boulder: Westview Press.

Gasparini, G. (1995). 'On Waiting.' *Time & Society* 4(1): 29–45.

Ghannam, F. (2013). *Live and Die Like a Man. Gender Dynamics in Urban Egypt*. Stanford: Stanford University Press.

Giddens, A. (1990). *The Consequences of Modernity*. Stanford: Stanford University Press.

Gilsenan, M. (1976). 'Lying, Honour, and Contradiction.' In B. Kapferer (ed.). *Transaction and Meaning: Directions in the Anthropology of Exchange and Symbolic Behavior*. Philadelphia: Institute for the Study of Human Issues: 191–219.

Gilsenan, M. (1996). *Lords of the Lebanese Marches: Violence and Narrative in an Arab Society*. Berkeley: University of California Press.

Graeber, D. (2011). 'Consumption.' *Current Anthropology* 52(4): 489–511.

Granqvist, H. (1935). 'Marriage Conditions in a Palestinian Village.' *Helsingfors: Societas Scientiarum Fennica, Commentationes Humanarum Litterarum* 6: 8.

Green, S.F. (2005). *Notes from the Balkans: Locating Marginality and Ambiguity on the Greek-Albanian Border*. Princeton: Princeton University Press.

Green, S. (2012). 'A Sense of the Border.' In T.M. Wilson and H. Donnan (eds.). *A Companion to Border Studies*. Oxford: Blackwell: 573–92.

Griffith, D. (2009). 'The Moral Economy of Tobacco.' *American Anthropologist* 111(4): 432–442.

Gullestad, M. (1996). 'From Obedience to Negotiation: Dilemmas in the Transmission of Values Between the Generations in Norway.' *The Journal of the Royal Anthropological Institute* 2(1): 25–42.

Gupta, A. (2005). 'Narratives of Corruption: Anthropological and Fictional Accounts of the Indian State.' *Ethnography* 6(1): 5–34.

Guyer, J. (1981). 'Household and Community in African Studies.' *African Studies Review*, Vol. 24(2/3): 87–137.

Guyer, J. and P. Peters (1987). 'Introduction.' *Development and Change*. Special Issue: Conceptualizing the Household. 18(2): 197–214.

Hamadeh, S. (2002). 'Feeding Calendar and Grazing Survey and Development of Rangeland Management Options for Target Areas in Northern Beka'a.' Project on Conservation and Sustainable Use of Sustainable and Agrobiodiversity of the Near East, Lebanon. http://www.lari.gov.lb/agrobio/Report%20final.pdf.

Hamadeh, S., M. Haidar and R. Zuraik (2006). *Research for Development in the Dry Arab Region: The Cactus Flower*. Southbound Penang: International Development Research Centre.

Hamadeh, S., R. Zurayk, F. Al Awar, S. Talhouk, D. Abi Ghanem, and M. Abi Said (1999). 'Farming System Analysis of Drylands Agriculture in Lebanon: An Analysis of Sustainability.' *Journal of Sustainable Agriculture* (15): 33–43.

Hamzeh, A. (1999). 'The Role of Islamic Movements in Municipal Elections [Dawr al-harakāt al Islāmiah fī al-intikhābāt al-baladiyya].' 'The Lebanese Municipal Elections of 1998: The Tribulations of Democracy within Local Communities.' *Lebanese Centre for Policy Studies* (Arabic version).

Hamzeh, A.N. (2000). 'Lebanon's Islamists and Local Politics: a New Reality.' *Third World Quarterly* 21(5): 739–759.

Hamzeh, A. (2001). 'Clientalism, Lebanon: Roots and Trends.' *Middle Eastern Studies* 37(3): 167–178.

Harik, J.P. (2004). *Hezbollah: The Changing Face of Terrorism*. London: I.B. Tauris.

Harb, M. (2010). 'On Religiosity and Spatiality: Lessons from Hezbollah in Beirut.' In N. AlSayyad and M. Massoumi (eds.). *The Fundamentalist City? Religiosity and the Remaking of Urban Space*. London and New York: Routledge.

Harris, M. (1998). 'The Rhythm of Life on the Amazon Floodplain: Seasonality and Sociality in a Riverine Village.' *Journal of the Royal Anthropological Institute* 4(1): 65–82.

Harris, O. (1981). 'Households as Natural Units.' In K. Young, C. Wolkowitz and R. McCullagh (eds.). *Of Marriage and the Market: Women's Subordination in International Perspective*. London: CSE Books: 49–68.

Hanf, T. (1993). *Coexistence in Wartime Lebanon. Decline of a State and Rise of a Nation*. London: The Centre for Lebanese Studies in association with I.B. Tauris & Co Ltd.

Hart, K. (2007). 'Love by Arrangement: the Ambiguity of "Spousal Choice" in a Turkish Village.' *The Journal of the Royal Anthropological Institute* 13(2): 345–362.

Hefner, R. (2000). *Civil Islam: Muslims and Democratization in Indonesia*. Princeton: Princeton University Press.

Henig, D. (2012). '"Knocking on my Neighbour's Door:" On Metamorphoses of Sociality in Rural Bosnia.' *Critique of Anthropology* 32(1): 3–19.

Hermez, S. (2015). 'When the State is (N)ever Present: On Cynicism and Political Mobilization in Lebanon.' *Journal of the Royal Anthropological Institute* 21(3): 507–523.

Hermez, S. (2017). *War is Coming: Between Past and Future Violence in Lebanon.* Philadelphia: University of Pennsylvania Press.

Hirsch, J.S., & Wardlow, H. (2006). *Modern Loves: The Anthropology of Romantic Courtship and Companionate Marriage.* Ann Arbor: University of Michigan Press.

Inhorn, M.C. (ed.) (2007). *Reproductive Disruptions: Gender, Technology, and Biopolitics in the New Millennium* (Vol. 11). Berghahn books.

Jabbra, J.G., and Nancy W. Jabbra (1978). 'Local Political Dynamics in Lebanon the Case of "Ain Al-Qasis."' *Anthropological Quarterly* 51(2): 137–151.

Jabbra, N.W. (2008). 'Family Change and Globalization in a Lebanese Village.' *International Journal of Sociology of the Family* 34(1): 65–76.

Jansen, S. (2014). 'Hope For/Against the State: Gridding in a Besieged Sarajevo Suburb.' *Ethnos* 79(2): 238–260.

Jansen, S. (2015). *Yearnings in the Meantime: 'Normal Lives' and the State in a Sarajevo Apartment Complex* (Vol. 15). Berghahn Books.

Jamieson, L. and R. Simpson (2013). *Living Alone: Globalization, Identity and Belonging.* Palgrave Macmillan.

Jean-Klein, I. (2000). 'Mothercraft, Statecraft, and Subjectivity in the Palestinian Intifada.' *American Ethnologist* 27(1): 100–127.

Johnson, M. (2001). *All Honourable Men: The Social Origins of War in Lebanon.* London, New York: I.B. Tauris.

Joseph, S. (1994). 'Brother/Sister Relationships: Connectivity, Love, and Power in the Reproduction of Patriarchy in Lebanon.' *American Ethnologist* 21(1): 50–73.

Joseph, S. (1996). 'Gender and Family in the Arab World.' In S. Sabbagh (ed.). *Arab Women between Defiance and Restraint.* New York: Olive Branch Press: 194–202.

Joseph, S. (1999). 'Descent of the Nation: Kinship and Citizenship in Lebanon.' *Citizenship Studies* 3(3): 295–318.

Joseph, S. (2005). 'Learning Desire: Relational Pedagogies and the Desiring Female Subject in Lebanon.' *Journal of Middle East Women's Studies* 1(1): 79–109.

Just, R. (1991). 'The Limits of Kinship.' In P. Loizos and E. Papataxiarchis (eds.). *Contested Identities. Gender and Kinship in Modern Greece.* Princeton, New Jersey: Princeton University Press: 114–134.

Kandiyoti, D. (1988). 'Bargaining with Patriarchy.' *Gender and Society* 2(3): 274–290.

Kandiyoti, D. (1995). 'Patterns of Patriarchy: Notes for an Analysis of Male Dominance in Turkish Society.' In Ş. Tekeli (eds.) *Women in Turkish Society.* London and New Brunswick, NJ: Zed Books Ltd.: 306–318.

BIBLIOGRAPHY

Kandiyoti, D. (1998). 'Gender, Power and Contestation: Rethinking Bargaining with Patriarchy.' In C. Jackson, and R. Pearson (eds.). *Feminist Visions of Development*. London: Routledge: 135–151.

Keddie, N. and B. Baron (eds.) (1991). *Women in Middle Eastern History. Shifting Boundaries in Sex and Gender*. New Haven: Yale University Press.

Khalaf, S. (1977). 'Changing Forms of Political Patronage in Lebanon.' In E. Gellner and J. Waterbury (eds.). *Patrons and Clients in Mediterranean Societies*. London: Duckworth.

Khlat, M. (1988). 'Consanguineous Marriage and Reproduction in Beirut, Lebanon.' *American Journal of Human Genetics* 43: 188–196.

Khlat, M. and A. Khudr (1984). 'Religious Endogamy and Consanguinity in Marriage Patterns in Beirut, Lebanon.' *Social Biology* 33: 138–145.

Kingston, P.W. (2013). *Reproducing Sectarianism: Advocacy Networks and the Politics of Civil Society in Postwar Lebanon*. New York: SUNY Press.

Kosmatopoulos, N. (2011). 'Toward an Anthropology of "State Failure:" Lebanon's Leviathan and Peace Expertise.' *Social Analysis* 55(3): 115–142.

Krayem, H. (2003). 'The Lebanese Civil War and the Taif Agreement.' *American University of Beirut* at http://ddc.aub.edu.lb/projects/pspa/conflict-resolution.html.

Kusserow, A.S. (1999). 'Crossing the Great Divide: Anthropological Theories of the Western Self.' *Journal of Anthropological Research* 55(4): 541–562.

Lamb, S. (2000). *White Saris and Sweet Mangoes. Aging, Gender and Body in North India*. California: University of California Press.

Lambek, M. (2011). 'Kinship as Gift and Theft: Acts of Succession in Mayotte and Israel.' *American Ethnologist* 38(1): 2–16.

Lambek, M. (2013). 'Kinship, Modernity, and the Immodern.' In S. McKinnon and F. Cannell (eds.). *Vital Relations: Modernity and the Persistent life of Kinship*. School for the Advanced Research: 241–60.

Lewinson, A.S. (2006). 'Domestic Realms, Social Bonds, and Class: Ideologies and Indigenizing Modernity in Dar Es Salaam, Tanzania.' *Canadian Journal of African Studies* 40(3): 462–495.

Li, T.M. (2005). 'Beyond "the State" and Failed Schemes.' *American Anthropologist* 107(3): 383–394.

Limbert, M. (2010). *In the Time of Oil: Piety, Memory, and Social Life in an Omani Town*. Stanford, CA: Stanford University Press.

Lipset, D. (2004). 'Modernity without Romance? Masculinity and Desire in Courtship Stories Told by Young Papua New Guinean Men.' *American Ethnologist* 31(2): 205–224.

Long, Nicholas J., and Henrietta L. Moore (2012). 'Sociality Revisited: Setting a New Agenda.' *The Cambridge Journal of Anthropology* 30(1): 40–47.

Maher, V. (1974). *Women and Property in Morocco. Their Changing Relation to the Process of Social Stratification in the Middle Atlas*. Cambridge: Cambridge University Press.

Marsden, M. (2005). *Living Islam: Muslim Religious Experience in Pakistan's North-West Frontier*. Cambridge: Cambridge University Press.

Marsden, M. (2007a). 'Love and Elopement in Northern Pakistan.' *The Journal of the Royal Anthropological Institute* 13(1): 91–108.

Marsden, M. (2007b). 'Islam, Political Authority and Emotion in Northern Pakistan.' *Contributions to Indian Sociology* 41(1): 41–80.

Marsden, T. (1998). 'Economic Perspectives.' In B. Ilbery (ed.). *The Geography of Rural Change*. London: Longman: 13–30.

Marx, E. (2006). 'The Political Economy of Middle Eastern and North African Pastoral Nomads.' In D. Chatty (ed.). *Nomadic Societies in the Middle East and North Africa. Entering the 21st Century* (Vol. 81). Leiden: Brill: 78–97.

McKinnon, S. and F. Cannell (2013). 'The Difference Kinship Makes.' In S. McKinnon and F. Cannell (eds.). *Vital Relations: Modernity and the Persistent life of Kinship*. School for the Advanced Research: 3–38.

Meier, D. (2013). 'The South Border: Drawing the Line in Shifting (Political) Sands.' *Mediterranean Politics* 18(3): 358–375.

Meillasoux, C. (1972). 'From Reproduction to Production. A Marxist Approach to Economic Anthropology.' *Economy and Society I*: 93–105.

Miller, D. (1997). 'Consumption and Its Consequences.' In H. Mackay (ed.). *Consumption and Everyday Life*. Sage: London: 1–50.

Miller, D. (ed.) (1998). *Material Cultures: Why Some Things Matter*. University of Chicago Press.

Mitchell, T. (1991). 'The Limits of the State: Beyond Statist Approaches and Their Critics.' *American Political Science Review* 85(1): 77–96.

Monroe, K.V. (2016). *The Insecure City*. New Jersey: Rutgers University Press.

Mouawad, J., and H. Bauman (2017). 'In Search of the Lebanese State.' *Arab Studies Journal* 25(1): 60–64.

Mormont, M. (1990). 'Who is Rural? Or, How to Be Rural: Towards a Sociology of the Rural.' In T. Marsden, P. Lowe and S. Whatmore (eds.). *Rural Restructuring. Global Processes and their Responses*. Fulton, London: 21–44.

Moser, C. (1993). *Gender Planning and Development*. London: Routledge.

Mughal, M.A.Z. (2015). 'Domestic Space and Socio-Spatial Relationships in Rural Pakistan.' *South Asia Research* 35(2): 214–234.

Mundy, M. (1995). *Domestic Government: Kinship, Community and Polity in North Yemen*. London, New York: I.B. Tauris.

Mundy, M., and B. Musallam (eds.) (2000). *The Transformation of Nomadic Society in the Arab East* (Vol. 58). Cambridge: Cambridge University Press.

Murad, M. (2004). *Baladiyyat Lubnan. Jadaliyyat al-tanmiya wa al-dimuqratiyya*. [Municipalities of Lebanon: Dialectics of Development and Democracy] Beirut: Dar Al-Mawasim.

Murray, C. (2001). 'Livelihoods Research: Some Conceptual and Methodological Issues.' Background paper 5. *Chronic Poverty Research Centre*. Department of Sociology, University of Manchester.

Nasr, S. (1978). 'Backdrop to Civil War: The Crisis of Lebanese Capitalism.' *Merip Reports* (73): 3–13.

Navaro-Yashin, Y. (2002). *Faces of the State: Secularism and Public Life in Turkey*. Princeton: Princeton University Press.

Nelson, M.K. (2006). 'Single Mothers "Do" Family.' *Journal of Marriage and Family* 68(4): 781–795.

Netting, R., R. Wilk and Eric Arnould (eds.). 1984. *Households: Comparative and Historical Studies of the Domestic Group*. Berkeley, Los Angeles, London: University of California Press.

Niamir-Fuller, M., and M.D. Turner (1999). 'A Review of Recent Literature on Pastoralism and Transhumance in Africa.' In M. Niamir-Fuller (ed.). *Managing Mobility in African Rangelands, the Legitimization of Transhumance*. London: Intermediate Technology Publications: 18–47.

Norton, A.R. (2007). *Hezbollah: A Short History*. New Jersey: Princeton University Press.

Nucho, J. (2016). *Everyday Sectarianism in Urban Lebanon. Infrastructures, Public Service, and Power*. New Jersey: Princeton University Press.

Nuijten, M.C.M. (2003). *Power, Community and the State: the Political Anthropology of Organisation in Mexico*. Pluto Press.

Obeid, M. (2006). 'Uncertain Livelihoods: Challenges Facing Herding in a Lebanese Village.' In Dawn Chatty (ed.). *Nomadic Societies in the Middle East and North Africa – Entering the 21st Century*. Leiden: Brill Publishers: 463–495.

Obeid, M. (2010). 'Development, Participation and Political Ideology in a Lebanese Town.' In S. Venkatesan and T. Yarrow (eds.). *Differentiating Development: Beyond an Anthropology of Critique*. Berghahn Books: 151–168.

Obeid, M. (2010a). 'Friendship, Kinship and Sociality in a Lebanese Town.' In A. Desai and E. Killick (eds.). *The Ways of Friendship: An Anthropological Exploration*. Oxford: Berghahn: 93–113.

Obeid, M. (2010b). 'Searching for the "Ideal Face of the State" in a Lebanese Border Town.' *Journal of the Royal Anthropological Institute* 16(2): 330–346.

Obeid, M. (2011). 'The "Trials and Errors" of Politics: Municipal Elections at the Lebanese Border.' *Political and Legal Anthropology Review* 34(2): 251–267. https://anthrosource.onlinelibrary.wiley.com/doi/10.1111/j.1555-2934.2011.01165.x.

Obeid, M. (2015). 'State of Aspiration: New Theoretical and Methodological Directions in the Middle East.' In S. Al-Torki (ed.). *The Anthropology of the Middle East*. Wiley-Blackwell Publishers: 434–451.

Pader, E.J. (1993). 'Spatiality and Social Change: Domestic Space Use in Mexico and the United States.' *American Ethnologist* 20(1): 114–137.

Paley, J. (2002). 'Toward an Anthropology of Democracy.' *Annual Review of Anthropology* 31(1): 469–496.

Peletz, M. (2001). 'Ambivalence in Kinship Since the 1940s.' In S. Franklin and S. McKinnon (eds.). *Relative Values: Reconfiguring Kinship Studies*. Durham and London: Duke University Press: 413–444.

Perthes, V. (1997). 'Myths and Money: Four Years of Hariri and Lebanon's Preparation for a New Middle East.' *Middle East Report* 203: 16–21.

Peters, E. (1980). 'The Status of Women in Four Middle East Communities.' In N. Kiddie and L. Beck (eds.). *Women in the Muslim World*. Cambridge: Harvard University Press: 311–350.

Picard, E. (2002). *Lebanon A Shattered Country. Myths and Realities of the Wars in Lebanon*. New York and London: Holmes & Meier.

Picard, E. (2006). 'Managing Identities Among Expatriate Businessmen Across the Syrian-Lebanese Boundary.' In I. Brandell (ed.). *State Frontiers. Borders and Boundaries in the Middle East*. London, New York: I.B. Tauris: 75–99.

Polanyi, K. (1957). 'The Economy as Instituted Process.' In K. Polanyi, C.M. Arensberg and H.W. Pearson (eds.). *Trade and Market in the Early Empires*. Glencoe: The Free Press: 243–70.

Rabo, A. (2008). '"Doing Family:" Two Cases in Contemporary Syria.' *Hawwa* 6(2): 129–153.

Radu, C. (2010). 'Beyond Border "Dwelling": Temporalizing the Border-Space through Events.' *Anthropological Theory* 10(4): 409–433.

Ramadan, A. (2009). 'Destroying Nahr el-Bared: Sovereignty and Urbicide in the Space of Exception.' *Political Geography* 28(3): 153–163.

Reed-Danahay, D. (2015). 'Social Space: Distance, Proximity, and Thresholds of Affinity.' In V. Amit (ed.). *Thinking Through Sociality: An Anthropological Interrogation of Key Concepts*. New York: Berghahn Books: 69–96.

Reeves, M. (2014). *Border Work: Spatial Lives of the State in Rural Central Asia*. Ithaca and London: Cornell University Press.

Rigg, J. (2014). *More than the Soil. Rural Change in Southeast Asia*. London and New York: Routledge.

Sa'ar, A. (2004). 'Many Ways of Becoming a Woman: The Case of Unmarried Israeli-Palestinian "Girls".' *Ethnology* 43(1): 1–18.

Sabean, D.W. (1990). *Property, Production and Family in Neckerhausen, 1700–1870*. Cambridge: Cambridge University Press.

Salibi, K. (1988). *A House of Many Mansions: The History of Lebanon Reconsidered*. London: I.B. Tauris.

Salloukh, B. (2005). 'Syria and Lebanon: A Brotherhood Transformed.' *Middle East Report* 236: 14–21.

Sayer, A. (2000). 'Moral Economy and Political Economy.' *Studies in Political Economy* 61(1): 79–103.

Schielke, S. (2009). 'Ambivalent Commitments: Troubles of Morality, Religiosity and Aspirations Among Young Egyptians.' *Journal of Religion in Africa* 39: 157–185.

Schneider, D. (1968). *American Kinship: A Cultural Account.* Chicago, IL: University of Chicago Press.

Schneidleder, A. (2017). 'Discreet and Hegemonic Borderscapes of Galilee: Lebanese Residents of Israel and the Israel–Lebanon Border.' *Geopolitics*: 1–22.

Schutz, A. (1967). *The Phenomenology of the Social World.* Evanston, IL: Northwestern University Press.

Scoones, I. (1994). *Living with Uncertainty: New Directions in Pastoral Development in Africa.* London: IT Publications.

Scoones, I. (2009). 'Livelihoods Perspectives and Rural Development.' *The Journal of Peasant Studies* 36(1): 171–196.

Scott, J. (1976). *The Moral Economy of the Peasant. Rebellion and Subsistence in Southeast Asia.* New Haven: Yale University Press.

Sharma, A. and A. Gupta (2006). 'Introduction: Rethinking Theories of the State in an Age of Globalisation.' In A. Sharma and A. Gupta (eds.). *The Anthropology of the State: A Reader.* Malden, Oxford, Victoria: Blackwell Publishing Ltd: 45–48.

Shehadeh, L.R. (2010). 'Gender-Relevant Legal Change in Lebanon.' *Feminist Formations* 22(3): 210–228.

Slyomovics, S., and S. Joseph (eds.) (2001). *Women and Power in the Middle East.* Philadelphia: University of Pennsylvania Press.

Spencer, J. (1997). 'Post-Colonialism and the Political Imagination.' *The Journal of the Royal Anthropological Institute* 3(1): 1–19.

Steinmetz, G. (ed.) (1999). *State/Culture: State-Formation After the Cultural Turn.* New York: Cornell University Press.

Tekçe, B. (2004). 'Paths of Marriage in Istanbul: Arranging Choices and Choice in Arrangements.' *Ethnography* 5(2): 173–201.

Thompson, E.P. (1971). 'The Moral Economy of the English Crowd in the Eighteenth Century.' *Past & Present* (50): 76–136.

Toren, C. (1999). 'Compassion for One Another: Constituting Kinship as Intentionality in Fiji.' *The Journal of the Royal Anthropological Institute* 5(2): 265–280.

Toren, C. (2012). 'Imagining the World that Warrants Our Imagination.' *The Cambridge Journal of Anthropology* 30(1): 64–79.

Traboulsi, F. (2007). *A History of Modern Lebanon.* London: Pluto Press.

Trouillot, M.R. (2001). 'The Anthropology of the State in the Age of Globalization.' *Current Anthropology* 42(1): 125–138.

Van Gelder, G.J. (2005). *Close Relationships. Incest and Inbreeding in Classical Arabic Literature*. London, New York: I.B. Tauris.

Volk, L. (2009). 'Martyrs at the Margins: The Politics of Neglect in Lebanon's Borderlands.' *Middle Eastern Studies* 45(2): 263–282.

Watts, M.J. (1992). 'Space for Everything (a Commentary).' *Cultural Anthropology* 7(1): 115–129.

Whitehead, A. (1981). 'I'm Hungry, Mum: The Politics of Domestic Budgeting.' In K. Young, C. Wolkovitz and R. McCullagh (eds.). *Of Marriage and the Market. Women's Subordination in International Perspective*. London: CSE Books: 93–116.

Wilk, R., R. Netting and E. Arnould (eds.) (1984). 'Households: Changing Forms and Functions.' *Households: Comparative and Historical Studies of the Domestic Group*. Berkeley, Los Angeles, London: University of California Press: 1–28.

Yan, Y. (2005). 'The Individual and Transformation of Bridewealth in Rural North China.' *The Journal of the Royal Anthropological Institute* 11: 637–658.

Yanagisako, S. (1984). 'Explicating Residence: A Cultural Analysis of Changing Households Among Japanese-Americans.' In R. Netting, R. Wilk and E. Arnould. *Households: Comparative and Historical Studies of the Domestic Group*. Berkeley, Los Angeles, London: University of California Press: 330–352.

Zigon, J. (2007). 'Moral Breakdown and the Ethical Demand: A Theoretical Framework for an Anthropology of Moralities.' *Anthropological Theory* 7: 131–150.

Index

Abrams, P.   49n8
Abu-Lughod, L.   16, 20, 90, 118n4, 119n6
Abu Rabia, A.   65, 66, 73, 75, 109, 119n6
affinity
   cross-border affinity   13
   marriage and   94
Agricultural Coordination Committee
   statement on unfair competition   48
agripastoral decline   41
agripastoral heritage, family life and   62–3
agripastoral labour organisation   22
Ahearn, L.   83, 84, 89
Al-Asad, Hafez   13, 13n8, 138
Al-Jamr, A.   49
Al-Nusra Front   154
Alatas, S.F.   123n1
Allerton, C.   110
American University of Beirut (AUB)   7
Amit, V., S. Anderson, V. Caputo, J. Postill, D. Reed-Danahay and G. Vargas-Cetina   19
Anderson, S.   36
Anti-Lebanon Mountains   8
anti-Syrian sentiments   148
'Arab Spring'   154
Arafat, Yaser   138
Aretxaga, B.   49n8
Arsal
   army involvement in electoral violence in   129
   Arsal List (Lā'ihat Arsal) in local elections   140, 141, 142, 144
   Arsal's Decision (Qarār Arsal) in local elections   140, 141, 144
   caricatured sketches of life in   11
   Civil War in Lebanon (1958), effects of   12–13
   Da'ish attack on, effects of   155–6
   demography of   21
   development needs   16–17
   economy of, Syrian control of   48–9
   fieldwork beginnings in   8, 9–10
   geography of   8–9
   growth in population   23
   lands of   21–2
   neighbourhoods of   7, 22–3, 32–3

   patronage in Beirut on behalf of, lack of   14–15
   'periphery of a periphery'   12
   physical expansion of   3
   state neglect following 1958 Civil War   12–13
   Sunni constituency in   15
   see also conversations of local people; memories
authority
   earlier generations and   84
   figures of   95, 97, 153
   household authority   86–7
   marriage and   100
   parental authority   89, 95, 100, 101
   personal authority   112
   state-provided authority, use of   144
awareness (wa'y)   124
   human development and   16

Baalbaki, A.   46, 64
Baalbek-Hermil Regional Development Programme   51
Barbour, B. and P. Salameh   83n1
Barlocco, F.   21, 25
Baroudi, S.E.   48
Baumann, H.   15
Baxter, D.   86, 107
Bell, S. and S. Coleman   21
Berliner D., M. Lambek, R. Shweder, R. Irvine, and A. Piette   115
Beydoun, A.   8n2
Biqā Northern Region   3
   alternative economies for, disappointments of   51
   Arsal in   9–10
   'bordering' in, Syrian crisis and new measures of   155
   character of   10
   farming protests in (2001)   48
   fruit farming, introduction of   4
   historical stereotype of   126
   Hizbullah social services in   15
   land use, changes in   4
   leadership (za'āma) in   128

marginality in   11
mobility patterns, changes in   4
multi-generational households in, lives in   7
normal lives, post-war yearning for   15
opium and cannabis production, attempts at eradication of   51
political landscape of, isolation in   153
political situation in, two-state rule and   41
rural lives, passing of time and changes in   4–5
rural society (*mujtama' rīfī*), rural modernities and   15–18
seasonality and movement of people in   65–6
Syrian presence in and harmony with   14
Syrian security (*mukhabarāt*) operations in   133
Bocco, R.   64n1
Bonte, P. and J.G. Galaty   63
Booth, W.J.   52
Borbieva, N.O.   100n7
border lives, predicament of   5–8, 148–9, 153–4, 156
Bourdieu, P.   30, 94
Brandell, I.   14n11
bribery by officialdom   134, 136–7
Brink-Danan, M.   146
brothers
    brother as trap (*al-akh fakh*)   107, 108, 114, 115
    enmity of   106–7
    sister/brother relationship   107–8
bureaucratic measures, seasonal movement and   64–5
Butt, B.   63
butter churning, task of   61

Cammett, M.   151n4
carpet workshop   7–8, 108–9, 112, 113, 114, 147
Carrithers, M.   20
Chamun, Kamil   12, 129
change
    cross-cultural trends in marriage and   83–4
    domestic governance, change in   80
    land use, changes in   4
    livelihoods, changes in   23

mobility patterns, changes in   4
modernity, social change and   15–18
national and regional influences on   6
physical and social environments, changes in   21–3
in rural livelihoods   41
rural lives, passing of time and changes in   4–5
textures of   6–7
transhumance, political change and   63–6
Chatty, D.   63, 66
checkpoints, deployment of   135–6
cherry farming   43–4
childhood adventures on Syrian-Lebanon border   44–5
choice (*khayār*) and marriage   82, 83, 89
Clarke, M.   85n3, 96n5, 122n7
clientalism, relationships of   15, 128
Coles, K.   125
Collier, J.F.   83, 83n2
Collier, J.F. and S.J. Yanagisako   106
colour schemes in domestic spaces   33–4
comfort in family life, ideal of   78–9
'comfortable woman' (*murtāha*)   63
    envy about   78–81
commensality, rejection of   122
communal lands
    expropriation of   44
    management of   64
connectivity
    connective relationships   107–8
    local and national   152–3
    notions of   116
    relational selves and   107
    technologies, connections and rise of   3
    virtues of   122
consent for marriage, necessity of   83, 89, 98, 100–101
consumer goods, gender identities and   39
consumption trends, social transformation and   81
conversation, gendered assumptions about   36
conversations with local people
    agripastoral heritage   62–3
    border lives, predicament of   5–8, 148–9, 153–4, 156
    brother/sister relationship   107–8

INDEX

childhood adventures on Syrian-Lebanon border   44–5
cooking collectively   31
council membership   135
crowded living, norm of   30–31
domestic space, social interactions and   33–4
domesticity, desire for ideals of   79
eating arrangements, food and   34
educational institutions   25
ethical governance, hopes for   142
evenings in winter   34–5
experimentation, use of vocabulary of   41–2
female competition   112
gender equality   85–6, 108
gender roles in social worlds   39
herding, cross-border pastures, bureaucracy and   65–6
herding, resources for, access to and control of   73–7
homemade or readymade (*khālis*) products   39
inheritance   108–9
interior furnishings, choice of   32–3, 34
intimidation, popular control and   133
Islamism, politics and   141, 144
kitchen arrangements   31–2
land, confiscation of   50
life events, sociality and   28–9
living lives, God and   92–3
local corruption   137–8, 140
marriage, practices and values in   83
marriage preferences of older people   88–9
modernity, betrayal, mistrust and   105–6
neighbourly relations   11, 12, 20, 23, 27–8, 29, 30, 111, 116–17, 152
'normal' lives, yearnings for   15, 130, 148–9, 153
partnership arrangements   66
pastoral returns, paucity of   54
physical and social environments, changes in   21–3
poverty and humble needs   2–3
privacy within households   31
quarrying, reflections on   56–7
representatives, mistrust of   137
romantic love and consent   89
sacrifice (*tadhiya*), kin loyalty and   108
social activities and reciprocity   26–7
socialization, inter-generational friendships and   35–6
spatial transformations and new relationships   25
state intervention, reflections on   51–2
Syrian officers, attitude towards   134
toilet facilities   31–2
transformation, lives in times of   5
transformation, politics and   145–6
see also memories
cooperation
   agricultural cooperation   28, 53
   cooperative herding   66
   ethos of (*ta'āwun*)   66, 80
   family, kinship and   54, 55–6, 63, 66, 73
   survival in pastoralism and   77
Coppolillo, P.B.   63
corruption   137–8, 140
   local elections and   128, 135–40, 142
Corsín Jiménez, A.   20
council membership   135
cousin marriage   86
Creed, G.W.   73
cross-cultural dowry practices   79
cross-sibling relationships   107
crowded living, norm of   30–31

Darwish, M.R., S. Hamadeh and M. Sharara   21, 44, 46–7
Darwish, T., C. Khater, I. Jomaa, R. Stehouwer, A. Shaban and M. Hamze   40
Das, V. and D. Pool   13
De Grazia, V., and E. Furlough   39
death, sociality in   28, 29
Deeb, L.   3–4n1, 16, 16n12, 78, 145n12
Deeb, L. and J. Winegar   18n13
Deek, C.   65
Delaney, C.   120
Democratic Alliance (*Al-Tahāluf al-Dimuqrāti*)   140, 141, 143, 144, 145, 147, 150, 152n5
'Democratic Choice' in local elections   130, 131
Dent, C.M.   114
desire (*raghbah*) and marriage   91, 95, 97, 101, 102
division of labour   68, 80
domestic architecture   31–2

domestic governance, change in   80
domestic spaces   30–35
　colour schemes   33–4
　consumer goods, gender identities and   39
　conversation, gendered assumptions about   36
　domestic architecture   31–2
　food preparation and consumption   34
　gendered sociality in   35–9
　generational mingling, value in   34–5
　heating   34
　house interiors   30–35
　interior furnishings   32–3
　living together, ethos of   33–4
　privacy, house interiors and   31, 33–4
　social interactions and   33–4
　spaciality within houses   33
　traditionalism and modernity in   34
　typical family pursuits in   34–5
domesticity   6, 32, 38
　desire for ideals of   79
　homemaking and, new language of   91
　role of women, change in   80–81
dowry practices   79
Duben, A. and C. Behar   95

eating arrangements, food and   34
educational institutions   25
Edwards, J. and M. Strathern   30
Eickelman, D. and J. Piscatori   144
Eisenhower Doctrine   129
El-Khazen, F.   13n7, 23, 131–2
El Nour S., C. Gharios, M. Mundy and R. Zurayk   16
El-Solh, R.   12n6
Elliot, A.   86
Ellis, F.   41
elopement   100–102
　variations of   101
engagements   79, 98
enmeshment, virtues of   122
'enslavement' in marriage, fear of   86
ethical governance, hopes for   142
ethnographic fieldwork   7–8
　scale and time, reconciliation of questions of   8
　see also methodological challenges
evenings in winter   34–5

everyday lives   3, 8, 27, 30, 110
　life events, sociality and   28–9
　living lives, God and   92–3
　living together, ethos of   33–4
　'living well,' contested matter of   52–3, 53–6
　political relations and   148
　rural lives, passing of time and changes in   4–5
　socio-economic and political influences on   6
　threats to   154
exchange marriage   88
experimentation (tajruba)
　elections as   125, 141
　in rural livelihoods   43–7
　use of vocabulary of   41–2

familism ('asabiyya 'ā'iliyya)
　dynamics of   127–8
　local elections and   123–4, 126–30, 132–5
　tensions driven by   128–9
family life
　agripastoral heritage and   62–3
　comfort in, ideal of   78–9
　'comfortable woman' (murtāha)   63
　commensality, rejection of   122
　consumption trends, social transformation and   81
　cross-cultural dowry practices   79
　denial or neglect of parents, shame in   121
　division of labour   68, 80
　domestic governance, change in   80
　domesticity and   79
　dynamics of, brother/sister dyad and   106–8
　engagements   79, 98
　entertainment of visitors   1, 33, 35, 81
　family resources, mother and distribution of   118–19
　gender and kinship relations   80–81
　genealogy of   19
　herding households   65–8, 69–71
　homemaking, women's role in   81
　Īd al Fitr, feast of   122
　inequalities in kinship structures, reproduction of   119–20
　labour needs and coherence of   62

INDEX 173

love and parental disapproval   61–2
motherhood, idolization of   121
neolocal residence, preferences for   79
open-mindedness (munfatih)   82, 98
personal dreams and aspirations   62
sanctity of family, cultural ideal of   71–2
social transformation, gender and kinship relations in   81
'talking' family and 'doing' family, distinction between   73
tensions between personal and family needs   62
unmarried women in   67, 75, 85–6, 107–8, 109–10, 111, 119
wilderness and sedentary lives compared   76
women, centrality in household activities   80–81
fate (*nasīb*)   103–4, 149
   linguistic use of   87
   marriage and   84–9
   resignation to   5, 59
   temporality and   86
female altruism   108–10
female competition   112
female vulnerability, ideas of   110–11
food preparation and consumption   34
forced marriage   87–9
Forte, T.   39
fruit farming   44
   competition problems for   48–9
   introduction of   4
Future Movement Party (*Al-Mustaqbal*)   148

Gasparini, G.   86
gathering together (*jam'a*), virtue in   28, 35
gender and kinship relations   80–81
gender equality   85–6, 108
gender roles in social worlds   39
gendered personhoods   6
gendered sociality in domestic spaces   35–9
gendered spaces and practices, construction of   36–7
genealogy of family life   19
generational comfort   3
generational mingling, value in   34–5
Ghannam, F.   115–16
Giddens, A.   83, 95
Gilsenan, M.   12, 30, 128

Graeber, D.   39
Granqvist, H.   109
Green, S.   8, 12, 155
Green, S.F.   149
Griffith, D.   52
Guevara, Che   23
Gullestad, M., M.   83
Gupta, A.   135
Guyer, J.   81
Guyer, J. and P. Peters   73

Hamadeh, S.   8n3, 62
Hamadeh, S., M. Haidar and R. Zuraik   8
Hamadeh, S., R. Zurayk, F. Al Awar, S. Talhouk, D. Abi Ghanem, and M. Abi Said   44, 47
Hamzeh, A.   125, 132
Hamzeh, A.N.   127, 145
Hanf, T.   12n6
Harb, M.   151n4
Harik, J.P.   145n12
Hariri, Rafiq   15, 16, 57
   assassination of, and aftermath of   15, 57, 146, 147, 150n3, 152–3, 154
Hariri, Saad   147, 148, 150–51, 152, 154
Harris, M.   27, 35
Harris, O.   73
Hart   95
Hart, K.   83, 84, 90
heating domestic spaces   34
Hefner, R.   145
Henig, D.   20, 20n1, 27n3
herding
   bureaucratic difficulties for   64–5
   collaborative labour of   54
   cross-border pastures, bureaucracy and   65–6
   dilemmas for   69–71
   gender considerations in   53–4
   herd monitoring, difficulties of   71–2
   Herders' Cooperative   7
   herding households   65–8, 69–71
   pastures in Syria for, accessibility of   65
   resources for, access to and control of   73–7
   seasonality of   67–8
   spatial divisions within households   67
   tradition of   53
   see also ownership of flocks

Hermez, S.   14, 48, 150
highlands (*jurd*)   1, 44–5, 76, 136
    desirability as habitat   56, 70
    health in work and   55
Hirsch, J.S. and H. Wardlow   83
Hizbullah   14, 147–8, 152
    anti-Hizbullah sentiments   57, 153
    Biqā Northern Region, political power in   151
    electoral strategy   145n12
    fighters in Syria   154
    Israel and, dealings with   141, 151–2
    pro-Hizbullah sentiments   151, 153
    social services in Biqā   15
homemade or readymade (*khālis*) products   39
homemaking   80–81, 91
    women's role in   81
Homs in Syria, frequency of trips to   10–11
hospitality   16, 27, 36–7, 122–3
    expectations of   17
households
    house interiors   30–35
    multi-generational households, lives in   7
    needs of, conflict of interest between personal lives and   71–8
    solidarity of, sheep as symbolic of   73
    see also family life
Hrawi, Ilyas   130n7
Hussein, A.   127–8
Hussein, Saddam   138

Ibn Khaldun   123n1
Īd al Fitr, feast of   122
individualism, marriage and   84
inequalities
    conflicts of interest in families and   73
    in kinship structures, reproduction of   108, 119–20
    regional inequalities   5, 15, 47
    structural inequalities of 'old times'   118–19
infrastructural deficiencies   10, 15, 135, 142, 153, 156
inheritance   108–10, 119
Inhorn, M.C.   85n3
inter-generational negotiations about marriage   84, 95–100, 102

inter-regional tensions   7, 151–2
interior furnishings   32–3
    choice of   34
interlineage marriage   99–100
intimidation, popular control and   133
Iraq War, protests about   138–40
Islamic Group (*Al-Jamāʿa al-Islāmiyyah*)   140, 142, 144–5, 146
Islamic State (Daʿish)   154
Islamism, politics and   141, 144

Jabbra, J.G. and N.W. Jabbra   128
Jabbra, N.W.   94n4
Jamieson, L. and R. Simpson   110
Jansen, S.   15, 148
Jean-Klein, I.   107
Johnson, M.   15, 128
Joseph, S.   15, 18, 84, 107–8, 125, 151n4
Joseph, S. and S. Slyomivics   120
Just, R.   106
Justice and Development (*Al-ʿAdāla wa al-Inmāʾ*)   140, 142, 144–5, 146

Kandiyoti, D.   109, 120
Karam, Najwa   61
Keddie, N. and B. Brown   120
Khalaf, S.   125
Khlat, M.   83n1
Khlat, M. and A. Khudr   83n1
kin marriage   82–3, 88
Kingston, P.W.   25
kinship
    brother as trap (*al-akh fakh*)   107, 108, 114, 115
    brothers, enmity of   106–7
    connective relationships   107–8
    connectivity, virtues of   122
    cross-sibling relationships   107
    enmeshment, virtues of   122
    failed bargains   115–20
    failure potential in   106
    family dynamics, brother/sister dyad and   106–8
    female altruism   108–10
    female vulnerability, ideas of   110–11
    genealogy of family life   19
    in herding   66–7
    kin relations, trust and   94–5
    kinship circles, marriage and   94–5

INDEX 175

local elections and challenge of   126–7
local value of   146
marriage and   86, 94
marriage and, creation and recreation of   94
morality of   105–6, 107, 120–22
patriarchy and   15, 84, 89, 107, 109, 116, 117–18, 120
replication of   106
rural livelihoods and   53
self-knowledge, reflexivity-in-action and   115–16
social relatedness, ontology of   111–12
spinsterhood   87, 110, 112
subjects related by   107–8
suspicion of kin, idioms of   107
unmarried women, roles of fathers and brothers in relation to   112–15
kitchen arrangements   31–2
Kosmatopoulos, N.   14
Krayem, H.   14n9
Kusserow, A.S.   84

Lahhud, Emile   128n5
Lamb, S.   116
Lambek, M.   53, 105, 117
lambing   69–70
land
    confiscation of   50
    land-use in Biqā, changes in   4
    land-use legislation, quarrying industry and   40–41, 42–3
    regulation of use of, idea of   50
    rents for, affordability of   65–6, 71
    suitability of, labour requirements and   70–71
    see also communal lands
Lebanon
    boundaries and thresholds, tensions and contradictions within   17
    changing times, War of 1975-1990 and border lives in   3–4, 5, 10, 11, 13
    Civil War in (1958)   12–13, 129
    Communist Party in, War of 1975-1990 and decline in popularity of   131–2
    educational institutions, social value of   25
    Environment, Ministry of   40–41
    Israeli Defence Forces (IDF) attacks on   151–2
    large-scale post-war reconstruction in   11–12
    livelihoods, War of 1975-1990 and experiments in   41, 44, 47, 50, 53
    marginalised areas (*muhammasha*) of   9–10
    National Master Plan (2004)   40–41, 51
    neighbourly conflict and War of 1975-1990 in   116–17
    pastoralism and War of 1975-1990 in   64, 77–8
    peripheries of, 'time warp' in   16
    political factionalism in, rise of   153–4
    rural and urban interconnections in   16–17
    'rural disintegration' in   5–6
    rural life in, cultural and moral milieu of   17–18
    'social coldness' of Hariri's development plans   15
    Syria and
        'Brotherhood, Cooperation and Coordination Agreement' between   14
        changing relationship between   6
        cross-border affinities   12–13
        seasonal movements of people between   7
    Ta'if Accord (Document of National Understanding)   13–14
    United Arab Republic project   12
    violence in, reverberations of   5–6
    War of 1975-1990 in   21, 23, 124, 141, 148, 150, 153
Lewinson, A.S.   30
Li, T.M.   49n8
Limbert, M.   3
lineage (*ā'ila*)
    clashes in marriage   99–100
    familism and, reconciliation with democracy   124–5
    local elections and   123–4, 125–6, 127, 130–32
    marriage and, honour and face of   88
Lipset, D.   84, 89
livelihoods
    changes in   23

competition problems for rural
   livelihoods   48–9
contested moral economies and rural
   livelihoods   52–60
diversification of   41, 42–3
family-based, transformation from   60
farming technologies, adoption of   28
multiple livelihood strategies   41–2, 46–7
War of 1975-1990 and experiments in   41,
   44, 47, 50, 53
local elections
   anti-Syrian politics, development of
      134–5
   anti-Syrian sentiments   133–4, 140
   Arsal List (*Lā'ihat Arsal*)   140, 141, 142, 144
   Arsal's Decision (*Qarār Arsal*)   140, 141, 144
   awareness (*wa'y*)   124
   bribery by Syrian officialdom   134, 136–7
   checkpoints, deployment of   135–6
   clientalism, relationships of   128
   Communist List in (1998)   131
   complicity of council in Iraq War
      protests   140
   'Consent' in   131
   corruption and   128, 135–40, 142
   council (2004), divisions within   144–5
   democracy, triumph in 1998 of   124
   Democratic Alliance (*Al-Tahāluf
      al-Dimuqrāti*)   140, 141, 143, 144, 145,
      147, 150, 152n5
   'Democratic Choice' in   130, 131
   destructive thuggery following   132–3
   election themes. endless rehearsal of
      124–5
   electoral lists (1998), determination of
      130–31
   electoral lists (2004)   140–44
   electoral lists (2004), determination of
      123–4
   experimentation (*tajruba*), elections as
      125, 141
   familism ('*asabiyya 'ā'iliyya*)   123–4,
      126–30, 132–5
   dynamics of   127–8
   tensions driven by   128–9
   idoim of lineage (*ā'ila*)   126
   Islamic Group (*Al-Jamā'a al-
      Islāmiyyah*)   140, 142, 144–5, 146
   Justice and Development (*Al-'Adāla wa
      al-Inmā'*)   140, 142, 144–5, 146

kinship, local value of   146
kinship and challenge of   126–7
leadership traditions and challenge of
   126–7
lineage (*ā'ila*)   123–4, 125–6, 127, 130–32
familism and, reconciliation with
   democracy   124–5
Lineage List in (1998)   131–2
lobbying for (2004)   123
local governance (*hukm mahallī*),
   negation of   135
mobility of local people, Syrian control
   over   135–6
modernity, 'backward' disruption of
   125–6
municipal work, non-sectarian honesty
   and seriousness in   145
national civil society campaign (1997)
   130
new council, new directions   144–6
'old wine in new bottles'   140–41, 142–4
passion in discussions about   123–4
patron-client relations   125
patronage, relationships of   128
political direction, need for integrity in
   142–4
political ideological affiliations   131–2
political representation, emergence of
   alternatives in   146
'politics of presence' in   146
quarrying industry, taxation problems
   and   136–7
satisfaction at outcome of 1998 election
   132
scheming in   127
secret services, influences of   129, 133–4
sectarian leadership   125
Syrian intelligence (*mukhabarāt*)
   involvement in politics   133–4, 135–6
transformatory power of   145–6
'Unity and Cooperation' in   131, 132
violence in   124, 127, 128–9, 130, 133
see also municipal elections
Long, N.J. and H.L. Moore   19
love
   desire and   90–92
   love marriage   77
   parental disapproval and   61–2

INDEX

Maher, V.   75, 119n6
marginality   5, 6, 18, 149
  in Biqā Northern Region   11
  border remoteness, marginal perspective and   8–12, 153–4
  marginalisation, sense of   15, 41, 47, 140
  marginalised areas (*muhammasha*) of Lebanon   9–10
  production process, women's marginalisation from   80–81
  Sunni population, marginalisation of   150–51
marriage
  affinity and   94
  authority and   100
  change in, cross-cultural trends and   83–4
  choice (*khayār*) and   82, 83, 89
  consent for, necessity of   83, 89, 98, 100–101
  cousin marriage   86
  creation and recreation of kinship and   94
  desire (*raghbah*) and   91, 95, 97, 101, 102
  elopement, variations of   101
  elopement and   100–102
  'enslavement,' fear of   86
  exchange marriage   88
  fate (*nasīb*) and   84–9
  forced marriage   87–9
  homemaking and   81
  individualism and   84
  intergenerational negotiations about   84, 95–100, 102
  interlineage marriage   99–100
  kin marriage   82–3, 88
  kinship and   86, 94
  kinship circles and   94–5
  lineage and, honour and face of   88
  lineage clashes   99–100
  love and desire   90–92
  love marriage   77
  marriage contract (*katb al-kitāb*)   88
  marriageability, emotional angst of women on cusp of   86–7
  married women, pastoralism and powers of   76–7
  modern marriage, vocabulary of   89–95
  modernity and   83–4
  modernity and attitudes to   89
  older people, preferences of   88–9
  parental complicity in elopement   101–2
  partnership and understanding   92–5
  pastoralism and   74, 75–6
  practices and values in   83
  progressive approaches to   95–8
  rights in, protection of   98–9
  romantic love and   82, 84, 89
  selfishness in elopement   101
  social sanctioning of, necessity for   100–101
  sociality in   28–9
  suitability for, doubts about   98
  tradition in, persistence of   82–3
  understanding and   94–5
  wanting (*irādah*) and   83, 91, 97, 102
  womanhood and   85–6
Marsden, M.   4, 90, 101, 144
Marx, E.   64
McKinnon, S. and F. Cannell   18
medical services, scarcity of   10
Meier, D.   8n2
Meillasoux, C.   71
memories
  of agripastoral livelihoods on 'old days'   43–4
  civil war, memories and fears of   150
  Communist memories of heroism in civil war   141
  of cross-border herding, problems of   64–5
  family experiences, recollections of   117–20
  of household chores and women's work in 'old days'   118
  mischievousness of youth   1
  'old days,' reflection on   1–2
  of quarrying, resilience and resourcefulness in   56–7
methodological challenges   5–6
  time, long-term perspectives and problem of   6–7
  see also ethnographic fieldwork
Migdal, J.S.   8
militarised political parties, proliferation of   13
Miller, D.   39
Mitchell, T.   49n8, 146
mobility

mobile pastoralism, assumptions about 63
patterns of, changes in 4
Syrian control over 135–6
modern marriage, vocabulary of 89–95
modernity
'backward' disruption of 125–6
betrayal, mistrust and 105–6
marriage and 83–4, 89
rural livelihoods and 56
rural society (*mujtama' rīfī*), rural modernities and 15–18
social change and 15–18
Monroe, K.V. 48
morality of kinship 105–6, 107, 120–22
Mormont, M. 16
Moser, C. 80n5
Mother of Martyrs (*Umm al-Shuhada*) epitaph after Israeli invasion (1982) 13, 152
motherhood, idolization of 121
Mouawad, J. and H. Bauman 14
Mughal, M.A.Z. 33
Mundy, M. 78, 81, 94, 119n6
Mundy, M. and B. Musallam 63
municipal elections (*intikhābāt baladiyya*) 125
  1964 event 126–30
  1998 event 124, 130–35
  2004 event 123–4, 140–46
Murad, M. 125–6, 130n7
Murray, C. 41
Muslim Law (*Sharī'a Muhammadiyya*) 66, 75, 85, 97, 108–10

Nasr, S. 5, 47
Nasrallah, Sayyid Hasan 152
Nasser, Jamal Abdel 23
Navaro-Yashin, Y. 49n8
neighbourhoods 111, 116, 121, 145, 151–2
  of Arsal 7, 22–3, 32–3
  conversations within 36, 78–9
  lineages and 28–9
  moral obligation within 27, 29
  physical transformation of 155–6
  spatial reordering of 4
neighbourly relations 11, 12, 20, 23, 27–8, 29, 30, 111, 116–17, 152
  obligations towards neighbours 27
Nelson, M.K. 73n3

neoliberal policies and rural livelihoods 47–8
neolocal residence, preferences for 79
Netting, R., R. Wilk and E. Arnould 81
newspapers, late arrival of 11
Niamir-Fuller, M. and M.D. Turner 65
'normal' lives, yearnings for 15, 130, 148–9, 153
Norton, A.R. 14n9–10, 145n12
Nucho, J. 151n4, 153
Nuijten, M.C.M. 49n8

Obeid, M. 7, 12, 14, 16, 25, 48, 133, 138, 149
open-mindedness (*munfatih*) 82, 98
opium and cannabis production, attempts at eradication of 51
ownership of flocks
  co-requisites for 74–5
  women and 75–6

Pader, E.J. 33
Paley, J. 146
parental complicity in elopement 101–2
parental denial or neglect, shame in 121
partnership
  arrangements for 66
  understanding in marriage and 92–5
past, perception of present and reflections on 3
pastoralism 7
  agripastoral life 62
  bureaucratic measures, seasonal movement and 64–5
  butter churning, task of 61
  'comfortable woman' (*murtāha*), envy about 78–81
  communal lands, management of 64
  cooperation and survival in 77
  cooperative herding 66
  environment and, relationship between 54–5
  family security and 53, 54
  hierarchies of age, gender and experience in 74
  household needs, conflict of interest between personal lives and 71–8
  household solidarity, sheep as symbolic of 73
  kinship in herding 66–7

INDEX

labour availability and sustainability of   63
labour requirements, land suitability and   70–71
lambing   69–70
land rents, affordability of   65–6, 71
marriage and   74, 75–6
married women, powers of   76–7
mobile pastoralism, assumptions about   63
organisation of, spatial and human   66–8
ownership of flocks, co-requisites for   74–5
ownership of flocks, women and   75–6
partnership in   66
pastoral returns, paucity of   54
resource conditions, factors shaping   63
sedentary living in town, highland pastoralism compared with   76
sheep marking   66
shepherd remuneration   75
summer camps   67–8
sustainability of, problems for   63
tensions between personal and family needs   71, 73, 77–8
tent-houses   67, 68
transhumance, political change and   63–6
winter camps   68
see also herding
patriarchy   15, 84, 89, 107, 109, 116, 117–18, 120
patron-client relations   50, 125
patronage, relationships of   128, 131n8, 144, 148
Peletz, M.   105
personal perspectives
   see conversations with local people; memories
Perthes, V.   15
Peters, E.   75, 109, 119n6
physical and social environments, changes in   21–3
Picard, E.   8, 12n6, 13, 13n8, 14
Polanyi, K.   52
politics
   engagement with   23–4
   ideological affiliations   131–2
   landscape in Biqā of, isolation in   153
   party ties, family lineage and   24
   polarization of   152–3
   political direction, need for integrity in   142–4
   political flux, effects of   64
   'politics of presence' in local elections   146
   representation in, emergence of alternatives   146
   representatives in government, mistrust of   137
   see also local elections
post-Syrian time   149–54
post-war lives, documentation of   5
poverty, humble needs and   2–3
privacy, house interiors and   31, 33–4
progressive approaches to marriage   95–8
Prophet Muhammad   27, 152

quarrying industry
   architectural style and   57
   day-labourers in, experiences of   58–9, 60
   economic benefits of   57–8
   employment stability in   42
   environmental activism and opposition to   59
   illegal quarrying, anxieties about   50–51
   income stability and   58
   land-use legislation and   40–41, 42–3
   legal uncertainties about   40–41
   medical costs and   58–9
   prosperity or sentence of hard labour in   56–60
   reflections on   56–7
   risks in   58–9
   rural livelihoods and   44, 46–7
   stonemasonery and   56–7
   stratification of   58
   taxation problems and   136–7
   wealth creation and   58
   workers in, problems of   40–41, 49
Qusayr, Battle of   154

Rabo, A.   73
Radu, C.   8
Ramadan, A.   155n9
reciprocal engagement   26
   value in   26–30

reciprocity 20, 28, 53, 56
    social activities and 26–7
recollections
    see conversations with local people; memories
Reed-Danahay, D. 20
Reeves, M. 8, 12, 155
representatives in government, mistrust of 137
resource conditions, factors shaping 63
Rigg, J. 17
rights in marriage, protection of 98–9
romantic love 82, 84
    consent and 89
Rural Development NGO 7–8
    carpet weaving and 'carpet girls' 7–8
    female productivity, concerns about *mate* and 37–8
    'local facilitator' 7–8
    social hub 7
rural livelihoods
    Agricultural Coordination Committee statement on unfair competition 48
    change in 41
    cherry farming 43–4
    communal lands, expropriation of 44
    competition problems for 48–9
    contested moral economies and 52–60
    cultural and moral milieu of 17–18
    disintegration in Lebanon of 5–6
    environment and pastoralism, relationship between 54–5
    experimentation in 43–7
    family-based, transformation from 60
    family labour and commitment to pastoralism 54
    farming, tradition of 53
    fruit growing 44
    highlands (*jurd*), health in work and 55
    identification as 'rural' 16–17
    kinship and 53
    'living well,' contested matter of 52–3, 53–6
    modernity and 56
    multiple livelihood strategies 41–2, 46–7
    neoliberal policies and 47–8
    pastoralism and family security 53, 54
    quarrying 44, 46–7
    smuggling 44–6, 57
    social value of production 52–3
    state and, ambivalent relationship 47–52
    sustainability in, search for 52
    sustainable development and 54–5
    urban investment and 47–8
    vulnerabilities of 41–2, 47
    'wilderness is freedom' (*al-barriyya hurriyya*) 55–6
    see also herding

Sa'ar, A. 85–6
Sabean, D.W. 83
sacrifice (*tadhiya*), kin loyalty and 108
Sahhaf, Muhammad Said 34–5
Salem, P. 152n5
Salibi, K. 12n6
Salloukh, B. 133, 134n9
sanctity of family, cultural ideal of 71–2
Sayer, A. 52
Schielke, S. 121
Schneider, D. 120
Schneidleder, A. 8n2
Schutz, A. 20n1
Scoones, I. 41, 65
Scott, J. 52
seasonality 21, 41
    agricultural work 56, 57–8, 79
    of health 55
    Lebanon and Syria, seasonal movement between 7, 21, 63–4, 65–6
    seasonal exchanges, sociality and 53
    seasonal fruits 36, 44
    social life and 27–8
secret services, influences of 129, 133–4
sectarianism 15, 57, 146, 150–52, 153
    sectarian leadership 125
sedentary living in town, highland pastoralism compared with 23, 76
self-knowledge, reflexivity-in-action and 115–16
selfishness in elopement 101
semi-nomadic agripastoralism 8–9
Sharma, A. and A. Gupta 49n8
sheep marking 66
Shehadeh, L.R. 109n2
shepherd remuneration 75
Shihab, Fuad 129
Slyomivics, S. and Joseph, S. 120

INDEX

smuggling   41, 49, 57
  decline in   136
  of diesel across Lebanese/Syrian border   25, 152
  of drugs across Lebanese/Syrian border   113
  rural livelihoods and   44–6, 57
  of weapons from Syria   152
social activism   24, 25, 29
social change across generations   2, 5–6, 81
social relatedness, ontology of   111–12
social value of production   52–3
sociality   19–20, 21–2
  'consociality'   20
  evening social visits in (*ziyārāt*)   2, 11
  gendered sociality   35–9
  kin-centred sociality   22–3, 24–5
  life events and   28–9
  relationality, 'mutual engagement' and (*'ishra*)   20, 21, 25
  spheres of, widening of   21–5
  War of 1975-1990 and   23–4
  workings of *'ishra*   26–30
socialization, inter-generational friendships and   35–6
societal metamorphosis   3
spatial transformations   20–21
  new relationships and   25
spatiality   35, 54, 70
  within houses   33
  human organization and   66–8
  neighbourhoods, spatial reordering of   4
  of social relationships   20–21
  socio-spatial lives   15
  spatial development   29
  spatial organization   30
  spatio-temporality   8, 10, 20
Spencer, J.   125
spinsterhood   87, 110, 112
state
  influence on livelihoods, malign nature of   49–50
  intervention of, reflections on   51–2
  rural livelihoods and, ambivalent relationship between   47–52
  state presence (or absence)   10, 48–9
Steinmetz, G.   49n8
stonemasonery   56–7
suffering, claims of exceptionalism in   149

summer camps   67–8
suspicion of kin, idioms of   107
sustainability
  of pastoralism, problems for   63
  in rural livelihoods, search for   52
  sustainable development   54–5
Syria
  'Brotherhood, Cooperation and Coordination Agreement' between Lebanon and   14
  conflict in   154
  crisis in, effects of   154–6
  cross-border affinities with Lebanon   12–13
  Lebanon and, changing relationship between   6
  officers from, attitude towards   134
  post-war subservience to   149
  presence in Biqā and harmony with   14
  refugees from, numbers of   154
  reliance of people on   14
  seasonal movements of people between Lebanon and   7, 21, 63–4, 65–6
  state intelligence (*mukhabarāt*) involvement in politics   133–4, 135–6

Ta'if Accord (Document of National Understanding)   13–14, 45
Tekçe, B.   95
tent-houses   67, 68
Thompson, E.P.   52
time passing, long-term knowledge and   6
toilet facilities   31–2
Toren, C.   20, 90
Traboulsi, F.   12n6, 14
traditionalism
  in marriage, persistence of   82–3
  modernity in domestic spaces and   34
  traditional livelihoods and way of life   52
transformation
  decline of agripastorilism and   38–9
  lives in times of   5
  physical transformation of neighbourhoods   155–6
  politics and   145–6
transhumance   22
political change and   63–6
transport business   25

Trouillot, M.R. 49n8
Tufaily, Sobhi 14n10

United Nations (UN)
　High Commissioner for Refugees (UNHCR) 154
　Resolution 1559 (2004) 147
'Unity and Cooperation' in local elections 131, 132
unmarried women
　family life and 67, 75, 85–6, 107–8, 109–10, 111, 119
　roles of fathers and brothers in relation to 112–15
urban investment, rural livelihoods and 47–8

Van Gelder, G.J. 96n5
violence in local elections 124, 127, 128–9, 130, 133
Volk, L. 10, 13, 14, 16, 156
vulnerabilities of rural livelihoods 41–2, 47

wanting (*irādah*) and marriage 83, 91, 97, 102
Watts, M.J. 12
Whitehead, A. 73
wilderness
　sedentary lives compared with lives in 76
　'wilderness is freedom' (*al-barriyya hurriyya*) 55–6
Wilk, R. and R. Netting 77, 80n5
winter camps 68
womanhood
　household activities and 80–81
　marriage and 85–6
　see also family life; unmarried women
Women's Food Cooperative 7

Yan, Y. 83
Yanagisako, S. 73, 81

Zigon, J. 106, 121–2

Printed in the United States
By Bookmasters